T0222386

TOPICS IN MODERN PHYSICS
Solutions to Problems

TOPICS IN MODERN PHYSICS
Solutions to Problems

Paolo Amore
Universidad de Colima, Mexico

John Dirk Walecka
College of William and Mary, USA

 World Scientific

NEW JERSEY · LONDON · SINGAPORE · BEIJING · SHANGHAI · HONG KONG · TAIPEI · CHENNAI

Published by

World Scientific Publishing Co. Pte. Ltd.

5 Toh Tuck Link, Singapore 596224

USA office: 27 Warren Street, Suite 401-402, Hackensack, NJ 07601

UK office: 57 Shelton Street, Covent Garden, London WC2H 9HE

Library of Congress Cataloging-in-Publication Data
Amore, Paolo, 1968–   author.
   Topics in modern physics : solutions to problems / Paolo Amore (Universidad de Colima, Mexico), John Dirk Walecka (College of William and Mary, USA World Scientific).
      pages cm
   Includes bibliographical references and index.
   ISBN 978-9814618953 (pbk. : alk. paper) -- ISBN 9814618950 (pbk. : alk. paper)
   1. Physics--Problems, exercises, etc. I. Walecka, John Dirk, 1932–   author. II. Title.
   QC32.A556 2014
   530--dc23

                              2014029358

British Library Cataloguing-in-Publication Data
A catalogue record for this book is available from the British Library.

Copyright © 2015 by World Scientific Publishing Co. Pte. Ltd.

All rights reserved. This book, or parts thereof, may not be reproduced in any form or by any means, electronic or mechanical, including photocopying, recording or any information storage and retrieval system now known or to be invented, without written permission from the publisher.

For photocopying of material in this volume, please pay a copying fee through the Copyright Clearance Center, Inc., 222 Rosewood Drive, Danvers, MA 01923, USA. In this case permission to photocopy is not required from the publisher.

Printed in Singapore

Preface

A three-volume series on modern physics has now been published by World Scientific Publishing Company (WSPC): *Introduction to Modern Physics: Theoretical Foundations* [Walecka (2008)], *Advanced Modern Physics: Theoretical Foundations* [Walecka (2010)], and *Topics in Modern Physics: Theoretical Foundations* [Walecka (2013)]. These books, aimed at the very best students, provide an introduction to the physics of the twentieth century. There are numerous problems included in the texts, and although the steps required in solving the problems are clearly laid out, it is evidently advantageous to have prepared solutions available. This is useful for both students and instructors, in order to calibrate their own solutions and to see how someone else would go about solving the problems.

When Dr. K. K. Phua, Executive Chairman of WSPC, suggested the possibility of having a solutions manual prepared for the first book *Introduction to Modern Physics: Theoretical Foundations*, the two of us discussed it and decided we would like to take on the task together. Although draft solutions to the problems were available to the authors upon publication of the text, it does take a significant amount of time and effort to get the solutions into publishable form.

This solutions manual has now been published by WSPC as *Introduction to Modern Physics: Solutions to Problems* [Amore and Walecka (2013)]. Solutions are presented for all of the problems in the initial modern physics volume. The problems are first restated, and then the solutions given. We try, as far as possible, to stick to the notation in the text. We believe these solutions are clear and concise. In a few cases, some useful additional material, not required in the solutions, has been included; it is so indicated. The use of readily available mathematics programs for personal computers provides a powerful tool in problem solving and pursuing the implications of the solutions, particularly those programs available to integrate both linear and non-linear differential equations; we make frequent use of this tool.

The first two books in the modern physics series provide an essentially linear progression, and many important modern physics topics were necessarily omitted in that presentation. The purpose of the third volume *Topics in Modern Physics: Theoretical Foundations* is to branch out from the first two, and provide an analysis of several of those additional topics. After an introductory chapter, the Topics Book is divided into three parts:

Part 1—Quantum Mechanics: Analytic solutions to the Schrödinger equation are first developed for some basic systems. The analysis is then formalized, concluding with a set of postulates that encompass the theory of quantum mechanics;

Part 2—Applications of Quantum Mechanics: The following applications of the theory are then discussed: approximation methods for bound states, scattering theory, time-dependent perturbation theory, electromagnetic radiation and quantum electrodynamics;

Part 3—Relativistic Quantum Field Theory: This part represents an extension of the material in the second volume, and contains chapters on discrete symmetries, the Heisenberg picture, and the Feynman rules for QCD.

The Topics Book is designed so that the material up through Part 2 follows on naturally from that in *Introduction to Modern Physics: Theoretical Foundations.* Over 135 accessible problems enhance and extend the coverage in the Topics Book. As with our previous solutions manual, the present volume provides the solutions to the entire set of problems in the book *Topics in Modern Physics: Theoretical Foundations* in what we believe to be a clear and concise fashion. Familiarity with the material in *Advanced Modern Physics: Theoretical Foundations* is assumed for the solutions to the problems in Part 3.

We were again delighted when World Scientific Publishing Company, which had done an exceptional job with several previous books, showed enthusiasm for publishing this new one. We would like to thank Dr. K. K. Phua, Executive Chairman of World Scientific Publishing Company, and our editor Ms. Lakshmi Narayanan, for their help and support on this project.

We believe that the three volumes in this modern physics series, together with the two solutions manuals, provide a base from which the very best students can learn and understand modern physics. When finished, readers should have an elementary working knowledge in the principal areas of twentieth-century physics. We sincerely hope that students and instructors alike will both enjoy and benefit from these solutions manuals; we certainly enjoyed writing them.

January 20, 2014

Paolo Amore
Facultad de Ciencias
Universidad de Colima
Colima, Mexico

John Dirk Walecka
College of William and Mary
Williamsburg, Virginia

Contents

Chapter 1

Introduction

Our understanding of the physical world was revolutionized in the twentieth century—the era of "modern physics". A book on *Introduction to Modern Physics: Theoretical Foundations* has been published by World Scientific Publishing Company (WSPC) [Walecka (2008)]. The goal of this book (Vol. I), aimed at the very best students, is to expose the reader to the foundations and frontiers of today's physics. Typically, students have to wade through several courses to see many of these topics, and the goal is to give them some idea of where they are going, and how things fit together, as they go along. A second book in this modern physics series (Vol. II) has also been published by WSPC as *Advanced Modern Physics: Theoretical Foundations* [Walecka (2010)]. Those few results quoted without proof in Vol. I are derived here. Again, an important goal is to keep the entire coverage self-contained.

A third book in this series *Topics in Modern Physics: Theoretical Foundations* has recently appeared [Walecka (2013)]. The purpose of this third volume is to branch out and cover in some detail several modern physics topics omitted in the essentially linear progression in Vols. I and II.

All of the books in this modern physics series contain an extensive set of accessible problems that enhance and extend the coverage. Although the steps required in their solution, and answers to them, are clearly laid out in the problems, the publisher urged that a solutions manual be prepared. In a previous book [Amore and Walecka (2013)], we provided a solutions manual for the problems in Vol. I. The first two parts of the "Topics Book" form a natural extension of Vol. I, and require only the material in that volume. The present book is a solutions manual for the entire set of over 135 problems in the book *Topics in Modern Physics: Theoretical Foundations*. The third part of the Topics Book extends the material in Vol. II.

The first two parts of the Topics Book are on *Quantum Mechanics* and *Applications of Quantum Mechanics*. Chapter 2 is on *Solutions to the Schrödinger Equation*. Here the problems start with the interpretation of the theory through the continuity equation and expectation values. General properties are derived for the eigenvalues and eigenfunctions of an hermitian hamiltonian. One problem provides insight through numerical integration of the differential equation. Other problems deal with properties of the solutions for the harmonic oscillator and hydrogen atom. The role of the wronskian is examined. An alternate derivation of the angular momentum operators is studied. Square-well bound states are investigated numerically. The circular box and three-dimensional oscillator are solved analytically in the problems, the former of relevance to two-dimensional systems and the latter of particular importance for calculations in nuclear physics. It is the authors' view that insight is obtained through extensive plotting of wave functions and probability densities, and chapter 2 has problems in this regard for the one-dimensional oscillator, hydrogen atom, circular box, spherical box, and three-dimensional oscillator. The scattering problem for a finite square barrier is solved, and its limits used to understand behavior in the Kronig-Penney model, which is also solved numerically in another problem. This model provides an archetype for propagation in a periodic structure. The time-dependent Schrödinger equation is solved in a three-dimensional box in the problems starting from a given initial configuration, and the probability density is then followed in time.

Chapter 3 of the Topics Book is on *Formal Developments*. After exercises on representations, completeness, and matrix mechanics, Ehrenfest's theorem is used to analyze the quantum oscillator. One problem derives the expansion of the finite operator transformation $e^{i\lambda\hat{A}}\hat{B}e^{-i\lambda\hat{A}}$ in repeated commutators, and another then uses this expansion to obtain the harmonic oscillator in the Heisenberg picture. There is a problem concerning a complete set of observables and measurements with a central force. Another problem demonstrates the invariance of the trace and determinant under unitary transformations.

Chapter 4 is on *Approximation Methods for Bound States*. One problem provides details of the calculation of the energy functional with a central force. Others prove that the energy functional always lies above the minimum value of the potential, and that the second-order energy shift is negative. A particularly important problem demonstrates that the first-order wave functions in non-degenerate perturbation theory form an orthonormal set. There are problems on second-order calculations of the energy-shift in the oscillator with λx^2 and γx^3 perturbations. A longer problem provides a variational calculation of the ground state of the H_2^+ molecular ion through the LCAO method, hence providing a description of the basic molecular bond. There are variational problems on the effect of a delta-function potential in a one-dimensional box, and on the ground state in a circular box. In connection with electromagnetic perturbations, another problem demonstrates the gauge-invariance of the Schrödinger equation. A three-dimensional plot is made of the dipole charge distributions in the $n = 2$ eigenstates for the Stark effect in hydrogen. The effect on the spectrum of hydrogen of the relativistic correction to the kinetic energy of the electron is calculated in perturbation theory. Another problem derives matrix mechanics from the variational principle.

Chapter 5 is on *Scattering Theory*. After exercises on the scattered wave, scattering wave function, and iteration of the scattering integral equation, the cross section for scattering from an exponential potential is evaluated in Born approximation. The optical theorem is derived from the scattering wave function. One problem provides an alternate path to the partial-wave decomposition of the scattering amplitude. Another uses numerical methods to obtain interesting results for the cross section for scattering from a hard sphere at all energies. There are problems which find the effective range and scattering length for square-well and square-well–hard-core potentials with a zero-energy bound state; additional problems derive these same results from effective-range theory. The behavior of the wave function at the turning point in the WKB approximation is studied, and the relation of the Glauber approximation to the Born approximation examined. Babinet's principle is derived in the eikonal approximation, and a simpler derivation of the optical theorem is obtained through the partial-wave expansion. Still another problem shows the relation of the S-matrix to the solutions to the partial-wave scattering equations with incoming- and outgoing-wave boundary conditions. An interesting exercise compares both the grey and black-disc eikonal models with high-energy p-p scattering. Finally, the classical attenuation of a beam is used to derive the leading expression for the optical potential in the eikonal approximation.

Chapter 6 is on *Time-Dependent Perturbation Theory*. The first problem concerns a numerical calculation of the Bessel function of imaginary argument involved in the energy-loss calculation in the text. Problems follow on the hydrogen atom in a discharging condenser and on the effective uncertainty relation $\Delta E \Delta t \sim h$. The differential and integrated cross sections for excitation of the $(2s, 2p)$ states in hydrogen through inelastic charged lepton $l^{\pm}$ scattering are obtained. Coulomb inelastic form factors are calculated and plotted for transitions to the $(3s, 3p, 3d)$ states. The cross section for ionization of hydrogen in charged lepton scattering demonstrates the extension of the analysis to two particles in the continuum in the final state. The cross section analysis is also extended to describe direct nuclear reactions, both in plane-wave and distorted-wave Born approximations. The probability current is generalized to include electromagnetic interactions and current conservation again established. The wavelength of radiation is compared with the size of the system for both nuclear and atomic radiative transitions. Finally, a problem focuses on the transition to the interaction picture and the role of the time-development operator $\hat{U}(t, t_0)$.

Chapter 7 is on *Electromagnetic Radiation and Quantum Electrodynamics*. The first problem demonstrates that there is enough flexibility in the normal-mode expansion of the vector potential $\mathbf{A}(\mathbf{x}, t)$ to match an arbitrary set of initial conditions. The expression for the momentum in the normal modes is derived. One problem shows that the integral of the Poynting vector over the surface of a small sphere and Gauss's theorem reproduce the expression for the energy-density in the free radiation field. Another presents an exercise on the equal-time commutation relations for the vector field operator. Other problems study the matrix element of the current in some detail. It is shown how the transverse delta-function projects the transverse part of a vector field. An uncertainty relation between field strength and photon number is derived, and it is shown how the quantum radiation field becomes a classical field. The free radiation field is put into the Heisenberg picture. A problem calculates the contribution of $1s \rightarrow 2p$ transitions to the dipole sum rule in hydrogen; the mean excitation energy implied by the sum rule is also obtained. It is shown that transitions one-level up exhaust the sum rule in the three-dimensional oscillator. There are three exercises on various aspects of the Wigner-Weisskopf theory of the line width.

The third part of the Topics Book is on *Relativistic Quantum Field Theory*, which extends the material in Vol. II. Chapter 8 is on *Discrete Symmetries*. The first problem justifies the form of the (π^+, π^-) state used in the text. The expressions for the Dirac spinors used in the symmetry discussion are then derived. Several relations used in the text are verified. An interesting problem demonstrates that the most general hermitian, Lorentz-invariant, isospin-invariant, and P-invariant Yukawa interaction between a nucleon and isovector pion is *separately* invariant under C and T. One problem investigates the implications for the scattering operator of the

.symmetry properties of the interaction hamiltonian density in the interaction picture. Another presents an application of time-reversal invariance for a specific scattering process. Implications of the anti-unitarity of the time-reversal operator are explored. A problem constructs the interaction hamiltonian density in the interaction picture from the lagrangian density used in the establishment of the CPT theorem in the text, and investigates its symmetry properties.

Chapter 9 in the Topics Book is on the *Heisenberg Picture*. There are problems on the adiabatic damping of the scattering operator $U(0, -\infty)$ and on its behavior under time reversal. The expressions used in the proof of the two basic theorems for the transition from the interaction picture to the Heisenberg picture are explicitly verified through order $\nu = 4$. The orthogonality of Heisenberg eigenstates belonging to distinct eigenvalues of the four-momentum operator is verified. An important problem deduces the general form of the matrix element of the electromagnetic current operator between single-nucleon states in the Heisenberg picture, and then investigates the consequences of P- and T-invariance for that matrix element.

Chapter 10 is on the *Feynman Rules for QCD*. One problem concerns the structure constants for SU(3). Another establishes various conservation laws from the equations of motion for QCD. Still another problem writes out those equations of motion in detail, as well as the Yukawa couplings in the theory. A more explicit, and very useful, treatment of indices is obtained by writing the quark wave functions in the quark field as the direct product of color, flavor, and Dirac wave functions. Gauss's law is obtained for QCD. The hamiltonian density of QCD is derived. The theory is then quantized in the Coulomb gauge. Finally, a problem based on the coupling of a Dirac particle with a scalar field provides an important exercise on deriving Feynman rules from the generating functional.

There are two appendices in the Topics Book on *The Two-Body Problem* and on a *Charged Particle in External Electromagnetic Field*. There are three problems on the transition to center-of-mass and relative coordinates in the former, and the proof of a vector relation used to establish the connection to Newton's law with the Lorentz force in the latter.

We believe the three volumes in this modern physics series, together with the solutions manuals, provide a clear, logical, self-contained, and comprehensive base from which the very best students can learn and

understand the subject.[1] When finished, readers should have an elementary working knowledge in the principal areas of theoretical physics of the
twentieth century. With this overview and these tools in hand, development
in depth and reach in this field can then be obtained from more detailed or
advanced physics courses and texts.

Familiarity with the material in the Topics Book provides a foundation
for practical applications of *quantum mechanics*, the marvelously success-
ful description we have of the microscopic world. There are indeed many
other good textbooks available with which readers can extend the depth
and breadth of their understanding of quantum mechanics, for example
[Schiff (1968); Messiah (1999); Gasiorowicz (2003); Griffiths (2004); Sakurai
and Napolitano (2010)]; at a more advanced level, there are [Dirac (1947);
Bjorken and Drell (1964); Feynman and Hibbs (1965); Landau and Lifshitz (1981); Shankar (1994); Merzbacher (1998); Gottfried and Yan (2004);
Cohen-Tannoudji, Diu, and Laloe (2006)].

---

[1]Particularly, when supplemented with [Walecka (2007)].

PART 1
Quantum Mechanics

Chapter 2

Solutions to the Schrödinger Equation

Problem 2.1 Assume the potential $V(\mathbf{x})$ is real, and show that the continuity Eq. (2.6) then follows from the Schrödinger Eq. (2.8).

Solution to Problem 2.1

Consider the three-dimensional Schrödinger equation

$$i\hbar\frac{\partial}{\partial t}\Psi(\mathbf{x},t) = \left[-\frac{\hbar^2}{2m}\nabla^2 + V(\mathbf{x})\right]\Psi(\mathbf{x},t)$$

where $V(\mathbf{x}) \equiv V(x,y,z)$ is a real potential. We interpret $|\Psi(\mathbf{x},t)|^2 d^3x$ as the probability of finding the particle in a small volume d^3x around $\mathbf{x}$, and we calculate

$$i\hbar\frac{\partial}{\partial t}\Psi^\star(\mathbf{x},t)\Psi(\mathbf{x},t) = \Psi^\star(\mathbf{x},t)\left[i\hbar\frac{\partial}{\partial t}\Psi(\mathbf{x},t)\right] + \left[i\hbar\frac{\partial}{\partial t}\Psi^\star(\mathbf{x},t)\right]\Psi(\mathbf{x},t)$$

With the use of the Schrödinger equation and its complex conjugate, this gives

$$i\hbar\frac{\partial}{\partial t}\Psi^\star(\mathbf{x},t)\Psi(\mathbf{x},t) = \Psi^\star(\mathbf{x},t)\left[-\frac{\hbar^2}{2m}\nabla^2 + V(\mathbf{x})\right]\Psi(\mathbf{x},t)$$
$$- \left\{\left[-\frac{\hbar^2}{2m}\nabla^2 + V(\mathbf{x})\right]\Psi^\star(\mathbf{x},t)\right\}\Psi(\mathbf{x},t)$$

where we have used the fact that V is real. The terms in V now cancel from this equation, and one is left with[1]

$$i\hbar\frac{\partial}{\partial t}\Psi^*(\mathbf{x},t)\Psi(\mathbf{x},t) = -\frac{\hbar^2}{2m}\left\{\Psi^*(\mathbf{x},t)\nabla^2\Psi(\mathbf{x},t) - [\nabla^2\Psi^*(\mathbf{x},t)]\Psi(\mathbf{x},t)\right\}$$

$$= -\frac{\hbar^2}{2m}\nabla\cdot\left\{\Psi^*(\mathbf{x},t)\nabla\Psi(\mathbf{x},t) - [\nabla\Psi^*(\mathbf{x},t)]\Psi(\mathbf{x},t)\right\}$$

This corresponds to the three-dimensional continuity equation

$$\frac{\partial\rho}{\partial t} + \nabla\cdot\mathbf{S} = 0$$

with

$$\rho = |\Psi|^2 \qquad\qquad \text{; probability density}$$

$$\mathbf{S} = \frac{\hbar}{2im}\left[\Psi^*\nabla\Psi - (\nabla\Psi^*)\Psi\right] \qquad \text{; probability current}$$

Problem 2.2 (a) Prove from the continuity equation that for a localized disturbance, the normalization integral is independent of time

$$\frac{d}{dt}\int\Psi^*\Psi\,d^3x = \frac{d}{dt}\int\rho\,d^3x = 0 \qquad \text{; localized disturbance}$$

(b) Use the Schrödinger equation to show quite generally that this condition holds if H is hermitian.

Solution to Problem 2.2

(a) With the use of the continuity equation, one obtains

$$\frac{d}{dt}\int\rho\,d^3x = \int\frac{\partial\rho}{\partial t}\,d^3x = -\int\nabla\cdot\mathbf{S}\,d^3x$$

Now use Gauss's theorem on this expression

$$\frac{d}{dt}\int\rho\,d^3x = -\int_A d\mathbf{A}\cdot\mathbf{S}$$

where A indicates a very large sphere surrounding the region of interest. If the disturbance is *localized*, then $\mathbf{S}$ vanishes on this surface, and the integral is zero.

$$\frac{d}{dt}\int\rho\,d^3x = 0$$

---

[1]Here we use the vector relation $\nabla\cdot(a\nabla b) = \nabla a\cdot\nabla b + a\nabla^2 b$.

(b) Use the Schrödinger equation to write

$$\frac{d}{dt} \int \rho \, d^3x = \int \frac{\partial(\Psi^\star \Psi)}{\partial t} \, d^3x = \int \left[\Psi^\star \frac{\partial \Psi}{\partial t} + \frac{\partial \Psi^\star}{\partial t} \Psi \right] d^3x$$

$$= \frac{1}{i\hbar} \int \left[\Psi^\star H\Psi - (H\Psi)^\star \Psi \right] d^3x$$

Now, quite generally, with any boundary conditions that leave the hamiltonian hermitian, this expression will vanish

$$\frac{d}{dt} \int \rho \, d^3x = 0$$

Problem 2.3 The expectation value of a physical quantity represented by the operator O is obtained in general from the following matrix element

$$\langle O \rangle = \frac{\langle \Psi(t)|O|\Psi(t) \rangle}{\langle \Psi|\Psi \rangle} \equiv \frac{\int \Psi^\star(\mathbf{x}, t) O \Psi(\mathbf{x}, t) \, d^3x}{\int \Psi^\star \Psi \, d^3x}$$

(a) Show the expectation value is real if O is hermitian;
(b) Show that for normal modes the energy is $\langle H \rangle = E;$[2]
(c) What is the particle's momentum in the coordinate representation?

Solution to Problem 2.3

(a) For an hermitian operator

$$\langle \Psi_a|O|\Psi_b \rangle = \langle \Psi_b|O|\Psi_a \rangle^\star$$

where (Ψ_a, Ψ_b) are any two acceptable wave functions. Hence

$$\langle O \rangle^\star = \langle O \rangle$$

and the expectation value of an hermitian operator is real.

(b) For normal modes $\Psi(t) = e^{-(i/\hbar)Et}$ where E is a real eigenvalue. The Schrödinger equation then gives

$$H\Psi = i\hbar \frac{\partial \Psi}{\partial t} = E\Psi$$

The expectation value of the hamiltonian is thus the eigenvalue E in normal modes

$$\langle H \rangle = E$$

---

[2] *Hint*: Use the Schrödinger equation.

(c) In the coordinate representation the particle's momentum is $\mathbf{p} = (\hbar/i)\nabla$, and one measures the expectation value. Therefore

$$\langle \mathbf{p} \rangle = \frac{\int \Psi^{\star}(\mathbf{x},t)(\hbar/i)\nabla\Psi(\mathbf{x},t)\,d^3x}{\int \Psi^{\star}\Psi\,d^3x}$$

Problem 2.4 Start from the eigenvalue Eq. (2.17) and its complex conjugate. Show

$$\int [\psi_n^{\star} H\psi_n - (H\psi_n)^{\star}\psi_n]d^3x = (E_n - E_n^{\star})\int |\psi_n|^2 d^3x$$

Now use the fact that H is hermitian to conclude the eigenvalues E_n must be *real*.

Solution to Problem 2.4

The eigenvalue equation and its complex conjugate are

$$H\psi_n = E_n\psi_n \qquad ; \ n = 0,1,2,\cdots,\infty$$
$$(H\psi_n)^{\star} = E_n^{\star}\psi_n^{\star}$$

Multiply the first equation by $\psi_n^{\star}$, the second by ψ_n, integrate over all space, and subtract. This gives

$$\int [\psi_n^{\star} H\psi_n - (H\psi_n)^{\star}\psi_n]d^3x = (E_n - E_n^{\star})\int |\psi_n|^2 d^3x$$

If the hamiltonian is hermitian, then the l.h.s. of the above expression vanishes, and thus

$$(E_n - E_n^{\star})\int |\psi_n|^2 d^3x = 0$$

Since the integral is positive definite, this relation gives

$$E_n = E_n^{\star}$$

Hence the eigenvalues are real.

Problem 2.5 Start from the eigenvalue Eq. (2.17) for ψ_n, and the complex conjugate equation for ψ_m. Show

$$\int [\psi_m^{\star} H\psi_n - (H\psi_m)^{\star}\psi_n]d^3x = (E_n - E_m^{\star})\int \psi_m^{\star}\psi_n\,d^3x$$

Use the fact that H is hermitian, and the result from Prob. 2.4, to conclude that the eigenfunctions corresponding to distinct eigenvalues are *orthogonal*.

Solution to Problem 2.5

The eigenvalue equation for ψ_n and the complex conjugate of the equation for ψ_m are

$$H\psi_n = E_n\psi_n \qquad ; n = 0, 1, 2, \cdots, \infty$$
$$(H\psi_m)^\star = E_m^\star\psi_m^\star$$

Multiply the first equation by $\psi_m^\star$, the second by ψ_n, integrate over all space, and subtract. This gives

$$\int [\psi_m^\star H\psi_n - (H\psi_m)^\star\psi_n]d^3x = (E_n - E_m^\star)\int \psi_m^\star\psi_n\, d^3x$$

If the hamiltonian is hermitian, then the l.h.s. of the above expression vanishes. Furthermore, from the previous problem the eigenvalues are real. Therefore

$$(E_n - E_m)\int \psi_m^\star\psi_n\, d^3x = 0$$

Since $E_n \neq E_m$, this relation gives

$$\int \psi_m^\star\psi_n\, d^3x = 0$$

Hence the eigenfunctions corresponding to distinct eigenvalues are orthogonal.

Problem 2.6 Suppose the eigenfunctions (ψ_1, ψ_2) correspond to the same eigenvalue. Introduce the linear combinations

$$\psi_1' \equiv \psi_1 \qquad ; \psi_2' \equiv \psi_2 - \left[\frac{\int \psi_1^\star\psi_2\, d^3y}{\int |\psi_1|^2 d^3y}\right]\psi_1$$

Show these degenerate wave functions are now *orthogonal*.[3]

Solution to Problem 2.6

It is clear that if

$$H\psi_1 = E\psi_1 \qquad ; H\psi_2 = E\psi_2$$

---

[3]This is the simplest example of the Schmidt orthogonalization procedure.

then

$$H\psi_1' = E\psi_1' \qquad ; \qquad H\psi_2' = E\psi_2'$$

Now by direct calculation

$$\int \psi_1'^\star \psi_2' \, d^3x = \int \psi_1^\star \psi_2 \, d^3x - \left[\frac{\int \psi_1^\star \psi_2 \, d^3y}{\int |\psi_1|^2 d^3y} \right] \int |\psi_1|^2 \, d^3x$$

$$= \int \psi_1^\star \psi_2 \, d^3x - \int \psi_1^\star \psi_2 \, d^3y$$

$$= 0$$

Hence these new degenerate wave functions are now orthogonal.

Problem 2.7 If the eigenfunctions $\psi_n(\mathbf{x})$ are complete, an arbitrary acceptable function $\phi(\mathbf{x})$ can be expanded as

$$\phi(\mathbf{x}) = \sum_{n=0}^{\infty} a_n \psi_n(\mathbf{x})$$

$$a_n = \int \psi_n^\star(\mathbf{y})\phi(\mathbf{y}) \, d^3y$$

where the last relation follows from the orthonormality of the eigenfunctions. Now justify the statement of completeness in Eq. (2.21).

Solution to Problem 2.7

Substitute the second expression into the first

$$\phi(\mathbf{x}) = \int d^3y \left[\sum_{n=0}^{\infty} \psi_n(\mathbf{x})\psi_n^\star(\mathbf{y}) \right] \phi(\mathbf{y})$$

The integral must reproduce $\phi(\mathbf{x})$ no matter what $\phi(\mathbf{y})$ does at all other points $\mathbf{y} \neq \mathbf{x}$. Hence the kernel in this integral must have the form

$$\sum_{n=0}^{\infty} \psi_n(\mathbf{x})\psi_n^\star(\mathbf{y}) = \delta^{(3)}(\mathbf{x} - \mathbf{y})$$

where the r.h.s. is a Dirac delta-function. This is the statement of completeness of the wave functions $\psi_n(\mathbf{x})$.

Problem 2.8 (a) Use Eq. (2.81) to generate $h_n(\xi)$ for $n = 1, 2, \cdots, 5$;
(b) Reproduce the results in (a) with Eq. (2.79), starting from $h_0(\xi) = 1$;

(c) Make a good plot of $(\hbar/m\omega_0)^{1/4}\psi_n$ and $(\hbar/m\omega_0)^{1/2}|\psi_n|^2$ as a function of ξ for the harmonic oscillator using these results.

Solution to Problem 2.8

(a) Equation (2.81) expresses the Hermite polynomials as

$$h_n(\xi) = (-1)^n e^{\xi^2} \frac{d^n}{d\xi^n} e^{-\xi^2}$$

This gives

$h_0(\xi) = 1$

$h_1(\xi) = -e^{\xi^2}\left(-2\xi e^{-\xi^2}\right) = 2\xi$

$h_2(\xi) = e^{\xi^2}\frac{d}{d\xi}\left(-2\xi e^{-\xi^2}\right) = 4\xi^2 - 2$

$h_3(\xi) = -e^{\xi^2}\frac{d^2}{d\xi^2}\left(-2\xi e^{-\xi^2}\right) = -e^{\xi^2}\frac{d}{d\xi}\left(-2e^{-\xi^2} + 4\xi^2 e^{-\xi^2}\right) = 8\xi^3 - 12\xi$

$\qquad\qquad\qquad\qquad\qquad\qquad\qquad\qquad\qquad\qquad\qquad ; etc.$

(b) Equation (2.79) is the recursion relation

$$h_{n+1}(\xi) = 2\xi h_n(\xi) - \frac{dh_n(\xi)}{d\xi}$$

With $h_0(\xi) = 1$, this provides a simpler way of arriving at the above results

$h_0(\xi) = 1$

$h_1(\xi) = 2\xi$

$h_2(\xi) = 4\xi^2 - 2$

$h_3(\xi) = 2\xi\left(4\xi^2 - 2\right) - 8\xi = 8\xi^3 - 12\xi \qquad ; etc.$

(c) The wave functions for the simple harmonic oscillator are given in Eqs. (2.92) as

$$\psi_n(x) = \left(\frac{m\omega_0}{\hbar}\right)^{1/4}\left(\frac{1}{2^n n!\sqrt{\pi}}\right)^{1/2} h_n(\xi)e^{-\xi^2/2} \qquad ; \xi = \left(\frac{m\omega_0}{\hbar}\right)^{1/2} x$$

The first four wave functions $\psi_n(x)$ and probability densities $|\psi_n(x)|^2$ are shown in Figs. 2.1 and 2.2 (taken from [Amore and Walecka (2013)]).

Problem 2.9 (a) Write, or obtain, a program to numerically integrate the differential Eq. (2.42) given an initial (ψ, ψ');

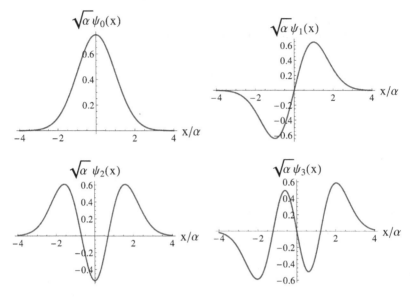

Fig. 2.1 First four wave functions of the one-dimensional simple harmonic oscillator. Here $\alpha \equiv (\hbar/m\omega_0)^{1/2}$.

(b) Start from $\xi \ll 0$ and the actual ground-state solution $\psi_0 \sim e^{-\xi^2/2}$. Chose an eigenvalue near, but not equal to, $\varepsilon_0 = 1$. What does your numerical solution look like for $\xi \gg 0$? Why? Show it is only for $\varepsilon_0 = 1$ that you can again match onto the decaying solution $\psi \sim e^{-\xi^2/2}$. Discuss;

(c) Repeat for the first excited state with $\psi_1 \sim \xi e^{-\xi^2/2}$ for $\xi \ll 0$.

Solution to Problem 2.9

(a) We want to investigate solutions to the differential Eq. (2.42)

$$\frac{d^2\psi}{d\xi^2} + (\varepsilon - \xi^2)\psi = 0$$

for given eigenvalues ε and initial conditions (ψ, ψ'). One can either write a simple numerical integration routine, or use any one of a number of available differential equation solvers, such as provided by Mathematica. In this case, the equation can actually be solved analytically with Mathematica. For instance, the command

```
solution[a_,b_,e_]:=DSolve[{Psi''[x]+(e-x^2)*Psi[x]==0,
         Psi[0]==a,Psi'[0]==b},Psi[x],x][[1]]
```

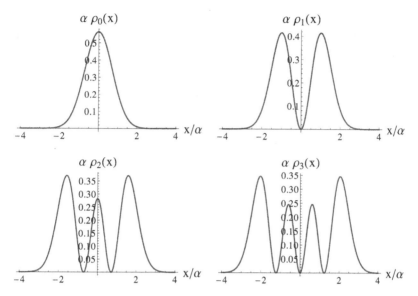

Fig. 2.2 Probability densities corresponding to the first four wave functions of the one-dimensional simple harmonic oscillator. Here $\rho_n = |\psi_n|^2$, and $\alpha \equiv (\hbar/m\omega_0)^{1/2}$.

solves Eq. (2.42) analytically, subject to the initial conditions $\psi(0) = a$ and $\psi'(0) = b$, corresponding to the eigenvalue $\varepsilon = $ e. This solution is expressed in terms of parabolic cylinder functions. Of course, in the case of differential equations that are more complicated than Eq. (2.42), an analytic solution may be out of reach. In these cases, as well as in the present case, one can resort to a purely numerical approach.

A simple modification of the previous command in Mathematica allows one to obtain the numerical solution to Eq. (2.42)

```
solution[a_,b_,e_,X_]:=NDSolve[{Psi''[x]+(e-x^2)*Psi[x]==0,
          Psi[0]==a,Psi'[0]==b},Psi[x],{x,0,X}][[1]]
```

In this case the numerical solution is found on the interval $[0, X]$, subject to the initial conditions $\psi(0) = a$ and $\psi'(0) = b$, corresponding to $\varepsilon = $ e. With the use of this command twice, once with $X = X_{\max} > 0$ and once with $X = X_{\min} < 0$, one obtains the numerical solution on an interval $[X_{\min}, X_{\max}]$, which contains the initial point $\xi = 0$.

Readers should notice that the choice of $\xi = 0$ as initial point is convenient here, since the solutions to Eq. (2.42) have definite parity, being either even or odd under the change $\xi \to -\xi$; therefore, one can easily isolate even and odd functions by imposing $\psi'(0) = 0$ and $\psi(0) = 0$, respectively.

(b) In Fig. 2.3 we plot the numerical solutions to Eq. (2.42) obtained with the command described in part (a), choosing $\psi(0) = 1$ and $\psi'(0) = 0$, for three different values of ε.

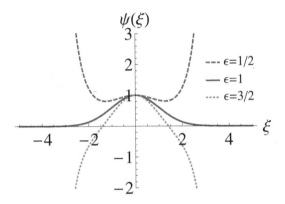

Fig. 2.3   Solutions to Eq. (2.42) choosing $\psi(0) = 1$ and $\psi'(0) = 0$, for three different values of ε.

As mentioned before, the exact solution to Eq. (2.42) may be found in terms of parabolic cylinder functions: in particular, choosing $\psi(0) = 1$ and $\psi'(0) = 0$ and setting $\varepsilon = 1 + \eta$, where $|\eta| \ll 1$, one may obtain the asymptotic behavior of $\psi(\xi)$ for $\xi \gg 1$ and $|\eta| \ll 1$, which reads[4]

$$\psi(\xi) \approx e^{-\xi^2/2} - \frac{\sqrt{\pi}\eta}{4\xi}\, e^{\xi^2/2} + \eta e^{-\xi^2/2} \left[\frac{1}{2}\log(2\xi) + \frac{i\pi}{4} + \frac{\gamma}{4} + \cdots\right] + \cdots$$

The sign of η determines the sign of the solution as $\xi \to \infty$. It is only for $\eta = 0$ that the second asymptotic solution $\sim e^{\xi^2/2}$ is not present, and the obtained solution dies off at infinity. This is clearly evidenced in Fig. 2.3.

(c) In Fig. 2.4 we plot the numerical solutions to Eq. (2.42) obtained with the commands described in part (a), choosing $\psi(0) = 0$ and $\psi'(0) = 1$, for three different values of ε.

In this case we may repeat the analytic analysis done in part (b), now with $\varepsilon = 3 + \eta$, considering separately the cases $\xi \to -\infty$ and $\xi \to \infty$:

---

[4]Here γ is Euler's gamma: $\gamma \approx 0.577216$.

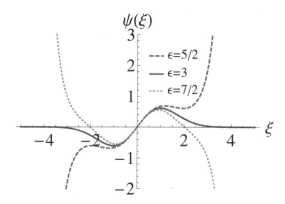

Fig. 2.4 Solutions to Eq. (2.42) choosing $\psi(0) = 0$ and $\psi'(0) = 1$, for three different values of ε.

For $\xi \to -\infty$ and $|\eta| \ll 1$ one has

$$\psi(\xi) \approx \xi e^{-\xi^2/2} + \frac{\sqrt{\pi}\eta}{8\xi^2} e^{\xi^2/2} +$$

$$\eta e^{-\xi^2/2} \left[\frac{1}{2}\xi \log(2\xi) + \frac{1}{4}(-2 + \gamma - i\pi)\xi - \frac{1}{8\xi} + \cdots \right] + \cdots$$

For $\xi \to \infty$ and $|\eta| \ll 1$ one has

$$\psi(\xi) \approx \xi e^{-\xi^2/2} - \frac{\sqrt{\pi}\eta}{8\xi^2} e^{\xi^2/2} +$$

$$\eta e^{-\xi^2/2} \left[\frac{1}{2}\xi \log(2\xi) + \frac{1}{4}(-2 + \gamma + i\pi)\xi - \frac{1}{8\xi} + \cdots \right] + \cdots$$

We see that the solutions blow up as $\xi \to \pm\infty$, with opposite signs in the two regions. For $\eta = 0$, corresponding to $\varepsilon = 3$, the second asymptotic solution $\sim e^{\xi^2/2}$ is absent, and the obtained solution to Eq. (2.42) is well-behaved and therefore physically acceptable. This is again clearly evidenced by the numerical solutions in Fig. 2.4.

Problem 2.10 Show the raising and lowering operators can be written

$$a^\dagger = \left(\frac{m\omega_0}{2\hbar}\right)^{1/2} \left(x - \frac{i}{m\omega_0}p\right) \qquad ; \; a = \left(\frac{m\omega_0}{2\hbar}\right)^{1/2} \left(x + \frac{i}{m\omega_0}p\right)$$

Solution to Problem 2.10

The raising and lowering operators are given in Eqs. (2.102) and (2.110) as

$$a^\dagger = \frac{1}{\sqrt{2}}\left(-\frac{d}{d\xi} + \xi\right) \qquad ; \ a = \frac{1}{\sqrt{2}}\left(\frac{d}{d\xi} + \xi\right)$$

With $\xi = (m\omega_0/\hbar)^{1/2}x$, this gives

$$a^\dagger = \left(\frac{m\omega_0}{2\hbar}\right)^{1/2}\left(x - \frac{\hbar}{m\omega_0}\frac{d}{dx}\right) \qquad ; \ a = \left(\frac{m\omega_0}{2\hbar}\right)^{1/2}\left(x + \frac{\hbar}{m\omega_0}\frac{d}{dx}\right)$$

Hence, with $p = (\hbar/i)d/dx$,

$$a^\dagger = \left(\frac{m\omega_0}{2\hbar}\right)^{1/2}\left(x - \frac{i}{m\omega_0}p\right) \qquad ; \ a = \left(\frac{m\omega_0}{2\hbar}\right)^{1/2}\left(x + \frac{i}{m\omega_0}p\right)$$

Problem 2.11 Show from the definition that the adjoint of a product is the product of the adjoints in the reverse order

$$(AB)^\dagger = B^\dagger A^\dagger$$

Solution to Problem 2.11

Given an operator O in quantum mechanics, the *adjoint* operator $O^\dagger$ is defined through the relation in Eq. (2.104)

$$\int (O\psi_l)^\star \psi_n \, dx \equiv \int \psi_l^\star O^\dagger \psi_n \, dx$$

which holds for any (ψ_l, ψ_n). Let O be the product AB. Then

$$\int (AB\psi_l)^\star \psi_n \, dx = \int \psi_l^\star (AB)^\dagger \psi_n \, dx$$

For A alone acting on the state $B\psi_l$

$$\int (AB\psi_l)^\star \psi_n \, dx \equiv \int (B\psi_l)^\star A^\dagger \psi_n \, dx$$

Now for B acting on the state $A^\dagger \psi_n$

$$\int (B\psi_l)^\star A^\dagger \psi_n \, dx = \int \psi_l^\star B^\dagger A^\dagger \psi_n \, dx$$

A comparison with the starting relation gives

$$(AB)^\dagger = B^\dagger A^\dagger$$

Problem 2.12 (a) Show a matrix element of the number operator satisfies

$$\langle\psi|a^\dagger a|\psi\rangle \equiv \int \psi^\star a^\dagger a\psi\,dx = \langle a\psi|a\psi\rangle \geq 0$$

(b) Use the result in (a) to prove that the eigenvalues of the hamiltonian in Eq. (2.125) are positive definite.

Solution to Problem 2.12

(a) From the definition of the adjoint

$$\int \psi^\star a^\dagger a\psi\,dx = \int (a\psi)^\star a\psi\,dx$$

Therefore

$$\langle\psi|a^\dagger a|\psi\rangle \equiv \int \psi^\star a^\dagger a\psi\,dx = \int (a\psi)^\star a\psi\,dx = \langle a\psi|a\psi\rangle \geq 0$$

Since $\langle\psi|\psi\rangle > 0$, it is then also true that

$$\frac{\langle\psi|a^\dagger a|\psi\rangle}{\langle\psi|\psi\rangle} \geq 0$$

(b) The eigenvalues ε of the hamiltonian in Eq. (2.125) are of the form

$$\left(2a^\dagger a + 1\right)\psi = \varepsilon\psi$$

Take the matrix element of this equation with ψ, and solve for ε

$$\varepsilon = \frac{\langle\psi|(2a^\dagger a + 1)|\psi\rangle}{\langle\psi|\psi\rangle} = 2\frac{\langle\psi|a^\dagger a|\psi\rangle}{\langle\psi|\psi\rangle} + 1$$

Now use the result in part (a) to obtain

$$\varepsilon \geq 1$$

Problem 2.13 (a) Use the commutation relation $[a, a^\dagger] = 1$, and the fact that $a\psi_0 = 0$, to show that

$$\langle\psi_0|aa^\dagger|\psi_0\rangle = 1$$

(b) Show through repeated use of the approach in (a) that

$$\langle\psi_0|a^n(a^\dagger)^n|\psi_0\rangle = n!$$

Hence establish the normalization of the eigenfunctions in Eq. (2.134).

Solution to Problem 2.13

(a) Use the commutation relation to write

$$aa^\dagger = [a, a^\dagger] + a^\dagger a = 1 + a^\dagger a$$

Since $a\psi_0 = 0$, the last term does not contribute to the matrix element. Hence

$$\langle \psi_0 | aa^\dagger | \psi_0 \rangle = 1$$

(b) Again, use the commutation relation to write

$$a(a^\dagger)^n = [a, a^\dagger]a^\dagger \cdots a^\dagger + a^\dagger [a, a^\dagger]a^\dagger \cdots a^\dagger + \cdots + a^\dagger \cdots a^\dagger [a, a^\dagger] + a^\dagger \cdots a^\dagger a$$
$$= n(a^\dagger)^{n-1} + (a^\dagger)^n a$$

Since $a\psi_0 = 0$, one finds for the matrix element

$$\langle \psi_0 | a^n (a^\dagger)^n | \psi_0 \rangle = \langle \psi_0 | a^{n-1} a (a^\dagger)^n | \psi_0 \rangle$$
$$= n \langle \psi_0 | a^{n-1} (a^\dagger)^{n-1} | \psi_0 \rangle$$

Now repeat this process n times

$$\langle \psi_0 | a^n (a^\dagger)^n | \psi_0 \rangle = n! \langle \psi_0 | \psi_0 \rangle = n!$$

This establishes the normalization of the eigenfunctions in Eq. (2.134).

Problem 2.14 Show the first few spherical harmonics are given by

$$Y_{00} = \frac{1}{\sqrt{4\pi}}$$

$$Y_{10} = \sqrt{\frac{3}{4\pi}} \cos\theta \qquad ; \ Y_{1,\pm 1} = \mp\sqrt{\frac{3}{8\pi}} \sin\theta \, e^{\pm i\phi}$$

$$Y_{20} = \sqrt{\frac{5}{16\pi}} (3\cos^2\theta - 1) \quad ; \ Y_{2,\pm 1} = \mp\sqrt{\frac{15}{8\pi}} \sin\theta \cos\theta \, e^{\pm i\phi}$$

$$; \ Y_{2,\pm 2} = \sqrt{\frac{15}{32\pi}} \sin^2\theta \, e^{\pm 2i\phi}$$

Solution to Problem 2.14

The spherical harmonics are defined in Eqs. (2.172) as

$$Y_{lm}(\theta, \phi) \equiv (-1)^m \left[\frac{2l+1}{4\pi} \frac{(l-m)!}{(l+m)!} \right]^{1/2} P_l^m(\cos\theta) e^{im\phi} \quad ; m \geq 0$$

$$Y_{lm}^* \equiv (-1)^m Y_{l,-m} \quad\quad\quad ; \text{defines others}$$

The Legendre and associated Legendre polynomials are in turn defined in Eqs. (2.170)

$$P_l^m(x) \equiv (1 - x^2)^{|m|/2} \frac{d^{|m|}}{dx^{|m|}} P_l(x)$$

$$P_l(x) = \frac{1}{2^l \, l!} \frac{d^l}{dx^l} (x^2 - 1)^l \quad\quad ; l = 0, 1, 2, \cdots$$

We then have

$$P_0(x) = P_0^0(x) = 1$$

$$P_1(x) = P_1^0(x) = x \quad\quad\quad ; P_1^1 = (1 - x^2)^{1/2}$$

$$P_2(x) = P_2^0(x) = \frac{1}{2}(3x^2 - 1) \quad ; P_2^1 = 3x(1 - x^2)^{1/2} \quad ; P_2^2 = 3(1 - x^2)$$

This implies for $l = 0$

$$Y_{00} = \frac{1}{\sqrt{4\pi}}$$

With $x = \cos\theta$, one finds for $l = 1$

$$Y_{10}(\theta, \phi) = \sqrt{\frac{3}{4\pi}} \cos\theta \quad ; Y_{1,\pm1}(\theta, \phi) = \mp\sqrt{\frac{3}{8\pi}} \sin\theta \, e^{\pm i\phi}$$

Finally, for $l = 2$

$$Y_{20}(\theta, \phi) = \sqrt{\frac{5}{16\pi}}(3\cos^2\theta - 1) \quad ; Y_{2,\pm1}(\theta, \phi) = \mp\sqrt{\frac{15}{8\pi}} \sin\theta \cos\theta \, e^{\pm i\phi}$$

$$; Y_{2,\pm2}(\theta, \phi) = \sqrt{\frac{15}{32\pi}} \sin^2\theta \, e^{\pm 2i\phi}$$

Problem 2.15 (a) Draw a more detailed picture, and provide a good geometric derivation of Eqs. (2.186) and (2.189);

(b) Derive Eqs. (2.187) and (2.190) algebraically through the transformation from cartesian to spherical coordinates

$$x = r\sin\theta\cos\phi \quad\quad ; y = r\sin\theta\sin\phi \quad\quad ; z = r\cos\theta$$

Solution to Problem 2.15

(a) In Fig. 2.5 we show the spherical coordinates (r, θ, ϕ) and associated orthogonal unit vectors $(\mathbf{e}_r, \mathbf{e}_\theta, \mathbf{e}_\phi)$. The equations appearing in (2.186) on the left can be understood looking at Fig. 2.6, which represents a section of Fig. 2.5 on the horizontal plane containing $\mathbf{e}_\phi$.

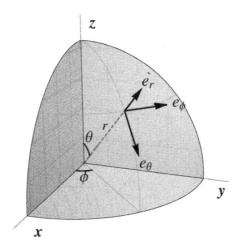

Fig. 2.5 Spherical coordinates (r, θ, ϕ) and associated orthogonal unit vectors $(\mathbf{e}_r, \mathbf{e}_\theta, \mathbf{e}_\phi)$.

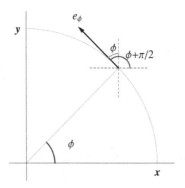

Fig. 2.6 Horizontal section of Fig. 2.5.

From this figure we see that $\mathbf{e}_\phi$ is perpendicular to $\mathbf{e}_z$ and that it forms angles of $\phi + \pi/2$ and ϕ with $\mathbf{e}_x$ and $\mathbf{e}_y$ respectively. This implies that $\mathbf{e}_x \cdot \mathbf{e}_\phi = \cos(\phi + \pi/2) = -\sin\phi$. Hence

$$\mathbf{e}_z \cdot \mathbf{e}_\phi = 0 \qquad ; \mathbf{e}_x \cdot \mathbf{e}_\phi = -\sin\phi \qquad ; \mathbf{e}_y \cdot \mathbf{e}_\phi = \cos\phi$$

We now look at Fig. 2.7 to understand the first of the equations appearing in (2.186) on the right.

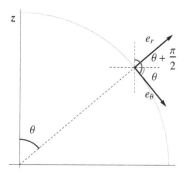

Fig. 2.7   Vertical section of Fig. 2.5.

We see that $\mathbf{e}_z$ and $\mathbf{e}_\theta$ form an angle $\theta + \pi/2$ and therefore $\mathbf{e}_z \cdot \mathbf{e}_\theta = \cos(\theta + \pi/2) = -\sin\theta$. To understand the last two equations we may observe that if $\mathbf{e}_\theta$ were lying in the plane generated by $\mathbf{e}_x$ and $\mathbf{e}_z$, it would form an angle θ with $\mathbf{e}_x$ and an angle $\pi/2$ with $\mathbf{e}_y$ (the reverse would clearly occur if we had chosen $\mathbf{e}_\theta$ lying in the plane generated by $\mathbf{e}_y$ and $\mathbf{e}_z$). If we now rotate $\mathbf{e}_\theta$ by an angle ϕ around the z-axis, we obtain the factors $\cos\phi$ and $\sin\phi$ in the last two equations. Thus

$$\mathbf{e}_z \cdot \mathbf{e}_\theta = -\sin\theta \qquad ; \mathbf{e}_x \cdot \mathbf{e}_\theta = \cos\theta\cos\phi \qquad ; \mathbf{e}_y \cdot \mathbf{e}_\theta = \cos\theta\sin\phi$$

(b) Consider the operator

$$L_z = \frac{1}{i}\left(x\frac{\partial}{\partial y} - y\frac{\partial}{\partial x}\right)$$

and apply it to a generic function $f(x, y, z) = \bar{f}(r, \theta, \phi)$. We first calculate the partial derivatives

$$\frac{\partial f}{\partial x} = \frac{\partial r}{\partial x}\frac{\partial \bar f}{\partial r} + \frac{\partial \theta}{\partial x}\frac{\partial \bar f}{\partial \theta} + \frac{\partial \phi}{\partial x}\frac{\partial \bar f}{\partial \phi}$$

$$= \sin\theta\cos\phi\frac{\partial \bar f}{\partial r} + \frac{\cos\theta\cos\phi}{r}\frac{\partial \bar f}{\partial \theta} - \frac{\sin\phi\csc\theta}{r}\frac{\partial \bar f}{\partial \phi}$$

$$\frac{\partial f}{\partial y} = \frac{\partial r}{\partial y}\frac{\partial \bar f}{\partial r} + \frac{\partial \theta}{\partial y}\frac{\partial \bar f}{\partial \theta} + \frac{\partial \phi}{\partial y}\frac{\partial \bar f}{\partial \phi}$$

$$= \sin\theta\sin\phi\frac{\partial \bar f}{\partial r} + \frac{\cos\theta\sin\phi}{r}\frac{\partial \bar f}{\partial \theta} + \frac{\cos\phi\csc\theta}{r}\frac{\partial \bar f}{\partial \phi}$$

$$\frac{\partial f}{\partial z} = \frac{\partial r}{\partial z}\frac{\partial \bar f}{\partial r} + \frac{\partial \theta}{\partial z}\frac{\partial \bar f}{\partial \theta} + \frac{\partial \phi}{\partial z}\frac{\partial \bar f}{\partial \phi}$$

$$= \cos\theta\frac{\partial \bar f}{\partial r} - \frac{\sin\theta}{r}\frac{\partial \bar f}{\partial \theta}$$

These results can now be used to obtain the explicit expressions for the components of the angular momentum operator; for example, with simple substitutions the third component of the angular momentum reads

$$L_z f(x,y,z) = \frac{1}{i}\left(x\frac{\partial f}{\partial y} - y\frac{\partial f}{\partial x}\right) = \frac{1}{i}\frac{\partial \bar f}{\partial \phi}$$

In a similar way, the remaining two of Eqs. (2.187) are also obtained

$$L_x f(x,y,z) = \frac{1}{i}\left(y\frac{\partial f}{\partial z} - z\frac{\partial f}{\partial y}\right) = \frac{1}{i}\left[-\sin\phi\frac{\partial \bar f}{\partial \theta} - \cos\phi\cot\theta\frac{\partial \bar f}{\partial \phi}\right]$$

$$L_y f(x,y,z) = \frac{1}{i}\left(z\frac{\partial f}{\partial x} - x\frac{\partial f}{\partial z}\right) = \frac{1}{i}\left[\cos\phi\frac{\partial \bar f}{\partial \theta} - \sin\phi\cot\theta\frac{\partial \bar f}{\partial \phi}\right]$$

In summary, we obtain Eqs. (2.187)

$$L_z = \frac{1}{i}\frac{\partial}{\partial \phi}$$

$$L_x = \frac{1}{i}\left[-\sin\phi\frac{\partial}{\partial \theta} - \cos\phi\cot\theta\frac{\partial}{\partial \phi}\right]$$

$$L_y = \frac{1}{i}\left[\cos\phi\frac{\partial}{\partial \theta} - \sin\phi\cot\theta\frac{\partial}{\partial \phi}\right]$$

Finally, to prove Eqs. (2.190) we need to explicitly calculate $L_i^2 f(x,y,z)$

for $i = x, y, z$. One has

$$L_x^2 f(x, y, z) = -\sin^2 \phi \frac{\partial^2 \bar{f}}{\partial \theta^2} - \cot \theta \left[\sin 2\phi \frac{\partial^2 \bar{f}}{\partial \theta \partial \phi} + \cos^2 \phi \left(\frac{\partial \bar{f}}{\partial \theta} + \cot \theta \frac{\partial^2 \bar{f}}{\partial \phi^2} \right) \right]$$
$$+ \frac{1}{4} \sin 2\phi (\cos 2\theta + 3) \csc^2 \theta \frac{\partial \bar{f}}{\partial \phi}$$

$$L_y^2 f(x, y, z) = -\cos^2 \phi \frac{\partial^2 \bar{f}}{\partial \theta^2} + \cot \theta \left[\sin 2\phi \frac{\partial^2 \bar{f}}{\partial \theta \partial \phi} - \sin^2 \phi \left(\frac{\partial \bar{f}}{\partial \theta} + \cot \theta \frac{\partial^2 \bar{f}}{\partial \phi^2} \right) \right]$$
$$- \frac{1}{4} \sin 2\phi (\cos 2\theta + 3) \csc^2 \theta \frac{\partial \bar{f}}{\partial \phi}$$

$$L_z^2 f(x, y, z) = -\frac{\partial^2 \bar{f}}{\partial \phi^2}$$

Therefore we obtain

$$\mathbf{L}^2 f(x, y, z) = -\frac{\partial^2 \bar{f}}{\partial \theta^2} - \cot \theta \frac{\partial \bar{f}}{\partial \theta} - \frac{1}{\sin^2 \theta} \frac{\partial^2 \bar{f}}{\partial \phi^2}$$

which is equivalent to Eq. (2.190)

$$\mathbf{L}^2 = - \left[\frac{1}{\sin \theta} \frac{\partial}{\partial \theta} \sin \theta \frac{\partial}{\partial \theta} + \frac{1}{\sin^2 \theta} \frac{\partial^2}{\partial \phi^2} \right]$$

Equation (2.190) can also be proven using Eqs. (2.188) in the text. In this case one needs to prove the relations contained in Eqs. (2.189). The simplest way to prove these equations is to use Eqs. (2.186) to express the unit vectors $(\mathbf{e}_r, \mathbf{e}_\theta, \mathbf{e}_\phi)$ in terms of the unit vectors $(\mathbf{e}_x, \mathbf{e}_y, \mathbf{e}_z)$.

We have

$$\mathbf{e}_\theta = (\mathbf{e}_x \cdot \mathbf{e}_\theta) \mathbf{e}_x + (\mathbf{e}_y \cdot \mathbf{e}_\theta) \mathbf{e}_y + (\mathbf{e}_z \cdot \mathbf{e}_\theta) \mathbf{e}_z$$
$$= \cos \theta \cos \phi \, \mathbf{e}_x + \cos \theta \sin \phi \, \mathbf{e}_y - \sin \theta \, \mathbf{e}_z$$
$$\mathbf{e}_\phi = (\mathbf{e}_x \cdot \mathbf{e}_\phi) \mathbf{e}_x + (\mathbf{e}_y \cdot \mathbf{e}_\phi) \mathbf{e}_y + (\mathbf{e}_z \cdot \mathbf{e}_\phi) \mathbf{e}_z$$
$$= -\sin \phi \, \mathbf{e}_x + \cos \phi \, \mathbf{e}_y$$

Moreover

$$\mathbf{e}_r = (\mathbf{e}_x \cdot \mathbf{e}_r) \mathbf{e}_x + (\mathbf{e}_y \cdot \mathbf{e}_r) \mathbf{e}_y + (\mathbf{e}_z \cdot \mathbf{e}_r) \mathbf{e}_z$$
$$= \sin \theta \cos \phi \, \mathbf{e}_x + \sin \theta \sin \phi \, \mathbf{e}_y + \cos \theta \, \mathbf{e}_z$$

Therefore

$$\frac{\partial \mathbf{e}_\phi}{\partial \theta} = 0$$

$$\frac{\partial \mathbf{e}_\phi}{\partial \phi} = -\cos\phi\,\mathbf{e}_x - \sin\phi\,\mathbf{e}_y = -\mathbf{e}_\theta\cos\theta - \mathbf{e}_r\sin\theta$$

$$\frac{\partial \mathbf{e}_\theta}{\partial \theta} = -\sin\theta\cos\phi\,\mathbf{e}_x - \sin\theta\sin\phi\,\mathbf{e}_y - \cos\theta\,\mathbf{e}_z = -\mathbf{e}_r$$

$$\frac{\partial \mathbf{e}_\theta}{\partial \phi} = -\cos\theta\sin\phi\,\mathbf{e}_x + \cos\theta\cos\phi\,\mathbf{e}_y = \mathbf{e}_\phi\cos\theta$$

This proves Eqs. (2.189).

Problem 2.16 Show that all three components of the angular momentum in Eqs. (2.187) are hermitian operators.

Solution to Problem 2.16

It is simpler to prove this in cartesian coordinates.[5] An operator is hermitian if it has the property

$$\langle \Psi_a | O | \Psi_b \rangle = \langle O \Psi_a | \Psi_b \rangle$$

Consider the operator L_x

$$\int d^3x\,\Psi_a^\star(\mathbf{x}, t) L_x \Psi_b(\mathbf{x}, t) = \frac{1}{i} \int d^3x\,\Psi_a^\star(\mathbf{x}, t) \left(y\frac{\partial}{\partial z} - z\frac{\partial}{\partial y} \right) \Psi_b(\mathbf{x}, t)$$

If this expression is integrated by parts, we arrive at

$$\int d^3x\,\Psi_a^\star(\mathbf{x}, t) L_x \Psi_b(\mathbf{x}, t)$$

$$= \frac{1}{i} \int dxdy\,[\Psi_a^\star(\mathbf{x}, t) y \Psi_b(\mathbf{x}, t)]_{z=-\infty}^{z=\infty} - \frac{1}{i} \int dxdz\,[\Psi_a^\star(\mathbf{x}, t) z \Psi_b(\mathbf{x}, t)]_{y=-\infty}^{y=\infty}$$

$$- \frac{1}{i} \int d^3x\,\Psi_b(\mathbf{x}, t) \left(y\frac{\partial}{\partial z}\Psi_a^\star(\mathbf{x}, t) - z\frac{\partial}{\partial y}\Psi_a^\star(\mathbf{x}, t) \right)$$

With the assumption that the wave functions vanish at infinite distances, the surface terms in the above expression cancel, and we are left with

$$\int d^3x\,\Psi_a^\star(\mathbf{x}, t)\,L_x\Psi_b(\mathbf{x}, t) = \int d^3x\,[L_x\Psi_a(\mathbf{x}, t)]^\star\,\Psi_b(\mathbf{x}, t)$$

Hence we conclude that with these boundary conditions, L_x is an hermitian operator. The proof for L_y and L_z is analogous.

---

[5]See [Amore and Walecka (2013)]; readers can investigate other boundary conditions.

Problem 2.17 Use the formulas for the spherical Bessel functions in [Schiff (1968)], or any other source, to verify the normalization constant in Eq. (2.222).

Solution to Problem 2.17

The solutions in a spherical box are given in Eqs. (2.221)

$$\psi_{nlm} = N_{nl} j_l \left(X_{nl} \frac{r}{a} \right) Y_{lm}(\theta, \phi) \qquad ; r \leq a$$

The normalization is then

$$1 = \int d^3r \, |\psi_{nlm}|^2 = |N_{nl}|^2 \int_0^a r^2 dr \, j_l^2 \left(X_{nl} \frac{r}{a} \right)$$

$$= |N_{nl}|^2 \left(\frac{a}{X_{nl}} \right)^3 \int_0^{X_{nl}} \rho^2 d\rho \, j_l^2(\rho)$$

Now use from [Schiff (1968)]

$$\int j_l^2(\rho) \rho^2 d\rho = \frac{1}{2} \rho^3 [j_l^2(\rho) - j_{l-1}(\rho) j_{l+1}(\rho)] \qquad ; l > 0$$

$$j_{l-1}(\rho) + j_{l+1}(\rho) = \frac{2l+1}{\rho} j_l(\rho) \qquad ; l > 0$$

For the spherical box, the wave function vanishes at the wall

$$j_l(X_{nl}) = 0$$

Therefore

$$|N_{nl}|^2 \left(\frac{a}{X_{nl}} \right)^3 \frac{1}{2} X_{nl}^3 \, j_{l+1}^2(X_{nl}) = 1$$

Hence, with an appropriate choice of phase,[6]

$$N_{nl} = \left[\frac{2}{a^3 j_{l+1}^2(X_{nl})} \right]^{1/2}$$

---

[6]The use of the following relation from [Schiff (1968)]

$$\int j_0^2(\rho) \rho^2 d\rho = \frac{1}{2} \rho^3 [j_0^2(\rho) + n_0(\rho) j_1(\rho)]$$

together with $j_0(X_{n0}) = 0$ and $n_0(X_{n0}) = j_1(X_{n0})$, shows that this result continues to hold for $l = 0$.

Problem 2.18 (a) Plot the radial probability densities for the states in a spherical box shown in Fig. 2.12 in the text;

(b) Use the results in Prob. 2.14 to make a polar plot of the probability densities for these states.

Solution to Problem 2.18

(a) The radial probability densities for the states in Fig. 2.12 in the text are easily plotted using Mathematica (or a similar math package). The spherical Bessel functions are related to the Bessel functions by the equation $j_l(x) = (\pi/2x)^{1/2} J_{l+1/2}(x)$. Since the zeros of $j_l(x)$ are also zeros of $J_{l+1/2}(x)$, one may use the built-in Mathematica command `BesselJZero[l+1/2,k]`, which produces the kth zero of $J_{l+1/2}(x)$, to obtain X_{nl} in Eq. (2.219). The radial probability density is then

$$P_{nlm}(r) = N_{nl}^2 \left[j_l \left(X_{nl} \frac{r}{a} \right) \right]^2 r^2$$

with N_{nl}^2 from Prob. 2.17. This probability density is plotted in Fig. 2.8.

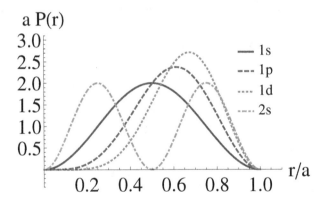

Fig. 2.8   Radial probability densities for the states in Fig. 2.12 in the text. Here $X_{10}/\pi = 1$, $X_{11}/\pi = 1.430$, $X_{21}/\pi = 1.835$, and $X_{20}/\pi = 2$ [Morse and Feshbach (1953)].

The full probability density is

$$\rho_{nlm}(r, \theta, \phi) = N_{nl}^2 \left[j_l \left(X_{nl} \frac{r}{a} \right) \right]^2 |Y_{lm}(\theta, \phi)|^2 r^2 \sin \theta$$

(b) We consider the angular probability density defined by

$$\Omega_{lm}(\theta, \phi) \equiv |Y_{lm}(\theta, \phi)|^2$$

where the spherical harmonics for the states in Fig. 2.8 are given in Prob. 2.14.

In Figs. 2.9, 2.10, and 2.11 we make a polar plot of the surface $\Omega_{lm}(\theta, \phi)$ [notice that $\Omega_{lm}(\theta, \phi)$ is really independent of ϕ, so that the corresponding plots are surfaces of revolution].

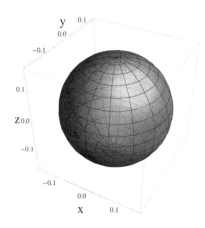

Fig. 2.9   Angular probability densities for the 1s- and 2s-states: $\Omega_{00}(\theta, \phi)$.

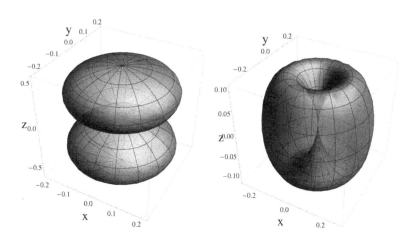

Fig. 2.10   Angular probability densities for the 1p-states: $\Omega_{10}(\theta, \phi)$ (left) and $\Omega_{1\pm1}(\theta, \phi)$ (right).

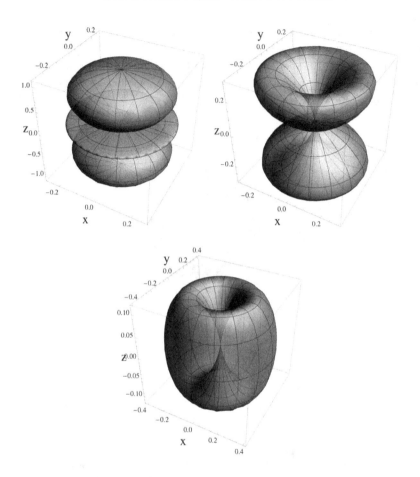

Fig. 2.11   Angular probability densities for the 1d-states: $\Omega_{20}(\theta, \phi)$ (left), $\Omega_{2\pm1}(\theta, \phi)$ (right) and $\Omega_{2\pm2}(\theta, \phi)$ (lower plot).

Problem 2.19 Two solutions $\{\psi_1, \psi_2\}$ to a second-order differential equation for $\psi(r)$ are said to form a *fundamental system* if their *wronksian* is non-zero

$$\psi_1\psi_2' - \psi_2\psi_1' \neq 0 \qquad ; \text{wronskian}$$

Show this is just the condition that allows one to match the initial value and slope $\{\psi, \psi'\}$ at an arbitrary point with a linear combination

$$\psi(r) = \alpha\psi_1(r) + \beta\psi_2(r)$$

Hence conclude that any solution can be expanded in terms of a fundamental system.

Solution to Problem 2.19

It is evident that if $\{\psi_1, \psi_2\}$ satisfy the linear second-order differential equation, then so does ψ. Suppose we try to specify the solution by matching the value and slope $\{\psi, \psi'\}$ at an arbitrary point r. Then

$$\alpha\psi_1(r) + \beta\psi_2(r) = \psi(r)$$
$$\alpha\psi_1'(r) + \beta\psi_2'(r) = \psi'(r)$$

Multiply the first equation by ψ_2', the second by ψ_2, and subtract. Solution for α then gives

$$\alpha = \frac{\psi(r)\psi_2'(r) - \psi'(r)\psi_2(r)}{\psi_1(r)\psi_2'(r) - \psi_2(r)\psi_1'(r)}$$

In a similar fashion, β is obtained as

$$\beta = \frac{\psi'(r)\psi_1(r) - \psi(r)\psi_1'(r)}{\psi_1(r)\psi_2'(r) - \psi_2(r)\psi_1'(r)}$$

The constants (α, β) are thus uniquely determined, provided the denominator, the *wronskian*, is non-zero

$$W(\psi_1, \psi_2) = \psi_1(r)\psi_2'(r) - \psi_2(r)\psi_1'(r) \neq 0$$

In this case, the solutions $\{\psi_1, \psi_2\}$ form a *fundamental system* in which any solution can be expanded.

Problem 2.20 (a) Discuss the graphical solution to the eigenvalue Eq. (2.229);

(b) Show the condition on the combination $V_0 a^2$ that there be just one bound state with $|E| \to 0$ is[7]

$$\frac{2m_0}{\hbar^2} V_0 a^2 = \frac{\pi^2}{4} \qquad ; \text{ zero-energy bound state}$$

(c) Sketch the corresponding wave function $u_0(r)$ in part (b).

---

[7]In the two-body problem, one must use the *reduced mass* $\mu = m_1 m_2/(m_1 + m_2)$ in this expression (see appendix A).

Solution to Problem 2.20

(a) First of all, we define the energy

$$\mathcal{V} \equiv \frac{\hbar^2 \pi^2}{8 m_0 a^2}$$

and re-cast Eq. (2.229) into the form

$$\frac{\pi}{2}\sqrt{\frac{V_0}{\mathcal{V}} - |\varepsilon|} \cot\left(\frac{\pi}{2}\sqrt{\frac{V_0}{\mathcal{V}} - |\varepsilon|}\right) = -\frac{\pi\sqrt{|\varepsilon|}}{2}$$

where $\varepsilon \equiv E/\mathcal{V}$. Clearly we are interested in solutions corresponding to a bound state, for which $-V_0/\mathcal{V} \leq \varepsilon \leq 0$: the graphical solution to this equation is illustrated in Fig. 2.12. The eigenvalues are given by the intersections of the two curves, which are the l.h.s. and r.h.s. of the above.

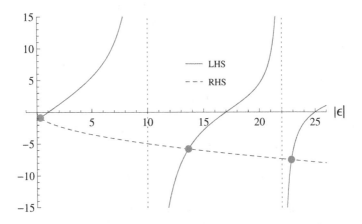

Fig. 2.12 Graphical solution to the eigenvalue equation for $V_0/\mathcal{V} = 26$.

(b) Consider the l.h.s. of Eq. (2.229) and take the limit $|E| \to 0$. In this case it reduces to

$$\text{l.h.s.} = \sqrt{\frac{2 m_0 a^2 V_0}{\hbar^2}} \cot\left(\sqrt{\frac{2 m_0 a^2 V_0}{\hbar^2}}\right)$$

In order to obtain a bound state with almost vanishing energy this expression must take a small negative value, which occurs when

$$\sqrt{\frac{2 m_0 a^2 V_0}{\hbar^2}} \to \frac{\pi}{2}$$

This condition is equivalent to

$$\frac{2m_0}{\hbar^2}V_0 a^2 = \frac{\pi^2}{4}$$

(c) The wave function $u_0(r)$ for $V_0 = 1.01\,V$ is plotted in Fig. 2.13 (notice that the wave function is not normalized).

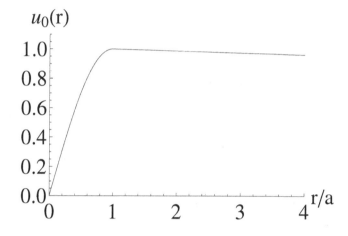

Fig. 2.13   Wave function $u_0(r)$ for $V_0 = 1.01\,V$

Problem 2.21 [Schiff (1968)] gives the following expression for the associated Laguerre polynomials

$$L_\kappa^q(\rho) = \sum_{\nu=0}^{\kappa-q} \frac{(-1)^{\nu+q}(\kappa!)^2 \rho^\nu}{(\kappa-q-\nu)!(q+\nu)!\nu!}$$

Start from Eqs. (2.277) and (2.271) and derive this result.

Solution to Problem 2.21

The definition of the associated Laguerre polynomial in Eq. (2.277), and the use of the Leibnitz formula for the multiple derivatives of a product in Eq. (2.271) give

$$L_\kappa^q(\rho) = e^\rho \frac{d^\kappa}{d\rho^\kappa}\left[\frac{\kappa!}{(\kappa-q)!}\rho^{\kappa-q}e^{-\rho}\right]$$

$$= \frac{\kappa!}{(\kappa-q)!}e^\rho \sum_{m=0}^{\kappa}\frac{\kappa!}{m!(\kappa-m)!}\left[\frac{d^m}{d\rho^m}\rho^{\kappa-q}\right]\left[\frac{d^{\kappa-m}}{d\rho^{\kappa-m}}e^{-\rho}\right]$$

Evaluation of the derivatives gives

$$L_\kappa^q(\rho) = \frac{(\kappa!)^2}{(\kappa-q)!} \sum_{m=0}^{\kappa-q} \frac{(-1)^{\kappa-m}}{m!(\kappa-m)!}(\kappa-q)(\kappa-q-1)\cdots(\kappa-q-m+1)\rho^{\kappa-q-m}$$

Now instead of summing over m, sum over

$$\nu \equiv \kappa - q - m$$

This gives

$$L_\kappa^q(\rho) = \frac{(\kappa!)^2}{(\kappa-q)!} \sum_{\nu=0}^{\kappa-q} \frac{(-1)^{q+\nu}}{(\kappa-q-\nu)!(\nu+q)!} \frac{(\kappa-q)!}{\nu!} \rho^\nu$$

Therefore

$$L_\kappa^q(\rho) = \sum_{\nu=0}^{\kappa-q} \frac{(-1)^{\nu+q}(\kappa!)^2\rho^\nu}{(\kappa-q-\nu)!(q+\nu)!\nu!}$$

Problem 2.22 Show the first three radial wave functions in the hydrogen atom are given by

$$R_{10}(r) = \left(\frac{Z}{a_0}\right)^{3/2} 2e^{-Zr/a_0}$$

$$R_{20}(r) = \left(\frac{Z}{2a_0}\right)^{3/2} \left(2 - \frac{Zr}{a_0}\right)e^{-Zr/2a_0}$$

$$R_{21}(r) = \left(\frac{Z}{2a_0}\right)^{3/2} \frac{Zr}{a_0\sqrt{3}}e^{-Zr/2a_0}$$

Solution to Problem 2.22

The radial wave functions for the hydrogen atom are given in Eqs. (2.283) as

$$R_{nl}(\rho) = \bar{N}_{nl}\,\rho^l e^{-\rho/2}L_{n+l}^{2l+1}(\rho) \qquad ; \rho = \frac{2Z}{n}\frac{r}{a_0}$$

$$\bar{N}_{nl} = N_{nl}(-1)^{2l+1}\frac{(n-l-1)!(2l+1)!}{[(n+l)!]^2}$$

where we adopt the phase convention that the N_{nl} are real and positive.[8]

---

[8]See [Schiff (1968)].

The absolute squares $|\bar{N}_{nl}|^2$ are given in Eqs. (2.295) as

$$|\bar{N}_{nl}|^2 = \left(\frac{2Z}{na_0}\right)^3 \frac{(n-l-1)!}{[(n+l)!]^3} \frac{1}{2n}$$

The associated Laguerre polynomials are defined in Eqs. (2.277)

$$L_\kappa^q(\rho) = e^\rho \frac{d^\kappa}{d\rho^\kappa} \left[\frac{\kappa!}{(\kappa-q)!} \rho^{\kappa-q} e^{-\rho}\right]$$

The three that we need are

$$L_1^1(\rho) = e^\rho \frac{d}{d\rho}\left[e^{-\rho}\right] = -1$$

$$L_2^1(\rho) = e^\rho \frac{d^2}{d\rho^2}\left[2\rho e^{-\rho}\right] = -2(2-\rho)$$

$$L_3^3(\rho) = e^\rho \frac{d^3}{d\rho^3}\left[6e^{-\rho}\right] = -6$$

The first radial wave function is therefore

$$R_{10}(r) = \left(\frac{2Z}{a_0}\right)^{3/2} \left(\frac{1}{2}\right)^{1/2} e^{-Zr/a_0} = \left(\frac{Z}{a_0}\right)^{3/2} 2\,e^{-Zr/a_0}$$

The second radial wave function is

$$R_{20}(r) = \left(\frac{Z}{2a_0}\right)^{3/2} \left(\frac{1}{2^2}\right)^{1/2} 2\left(2 - \frac{Zr}{a_0}\right) e^{-Zr/2a_0}$$
$$= \left(\frac{Z}{2a_0}\right)^{3/2} \left(2 - \frac{Zr}{a_0}\right) e^{-Zr/2a_0}$$

The third radial wave function is

$$R_{21}(r) = \left(\frac{Z}{2a_0}\right)^{3/2} \left(\frac{1}{3^3 2^2}\right)^{1/2} 6\left(\frac{Zr}{a_0}\right) e^{-Zr/2a_0}$$
$$= \left(\frac{Z}{2a_0}\right)^{3/2} \frac{Zr}{a_0\sqrt{3}} e^{-Zr/2a_0}$$

Problem 2.23 It is the s-states that get into the origin in the hydrogen atom. Start from Eq. (2.267) and compute the radial derivative $\partial\psi_{n00}(r)/\partial r|_{r=0}$. Make a three-dimensional sketch of $\psi_{n00}(r)$ for small r. Discuss.

Solution to Problem 2.23

We use Eq. (2.267), together with the normalization constant in Eq. (2.295), to obtain the wave functions for the s-states of the hydrogen–like atom

$$\psi_{n00}(r, \theta, \phi) = \frac{1}{\sqrt{\pi}} \left(\frac{Z}{na_0}\right)^{3/2} e^{-rZ/na_0} F(1 - n|2| 2zr/na_0)$$

It is easy to see from Eq. (2.255) that

$$\frac{d}{d\rho} F(a|b|\rho) = \frac{a}{b} F(1 + a|1 + b|\rho)$$

With the use of this property we have

$$\frac{\partial \psi_{n00}(r, \theta, \phi)}{\partial r} = -\frac{1}{\sqrt{\pi}} \left(\frac{Z}{na_0}\right)^{5/2} e^{-rZ/na_0} [F(1 - n|2| 2zr/na_0)$$
$$+ (n - 1) F(2 - n|3| 2zr/na_0)]$$

This derivative can now be evaluated at $r = 0$ with the aid of Eq. (2.296)

$$\frac{\partial \psi_{n00}(r, \theta, \phi)}{\partial r}\bigg|_{r=0} = -\frac{1}{\sqrt{\pi n^3}} \left(\frac{Z}{a_0}\right)^{5/2}$$

Thus we see that for the s-states of hydrogen-like atoms the wave function has a negative slope at the origin in the radial direction: for a fixed Z, the magnitude of this slope decreases as the principal quantum number increases.

In terms of the dimensionless variable $u \equiv Zr/a_0$, we consider the quantity

$$\left(\frac{a_0}{Z}\right)^{3/2} \psi_{n00}(r, \theta, \phi) = \sqrt{\frac{1}{\pi n^3}} e^{-u/n} F(1 - n|2| 2u/n)$$

In the vicinity of $u = 0$, we have

$$\left(\frac{a_0}{Z}\right)^{3/2} \psi_{n00}(r, \theta, \phi) \approx \frac{1 - u}{\sqrt{\pi n^3}} + O\left(u^2\right)$$

The scaled wave functions for the s-states of the hydrogen-like atom are plotted in Fig. 2.14: the dashed lines are the leading behavior for $r \to 0$.

Figure 2.15 is a three-dimensional plot of $(a_0/Z)^{3/2} \psi_{100}(r, \theta, \phi)$ in the vicinity of the origin, as viewed in the x-y plane. Note the cusp in the wave function at the origin that reflects the singular nature of the point Coulomb potential.

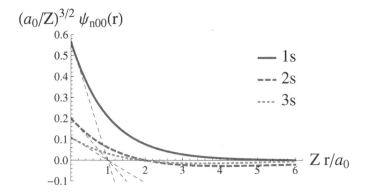

Fig. 2.14 Wave functions for the s-states in the vicinity of the origin.

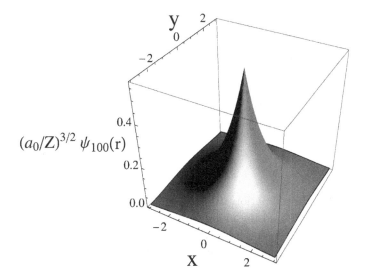

Fig. 2.15 Three-dimensional plot of $(a_0/Z)^{3/2} \psi_{100}(r, \theta, \phi)$ in the vicinity of the origin, as viewed in the x-y plane.

Problem 2.24 The capture of negative muons on nuclei takes place between a muon in a $1s$ atomic orbit and a proton in the nucleus.

(a) Use the Bohr model to demonstrate that for light nuclei, the muon orbit lies well inside the atomic electron cloud and outside of the nucleus;

(b) The reaction proceeds through the contact weak interaction

$$\mu^- + p \to \nu_\mu + n$$

Show that the capture-rate is proportional to Z^4 for light nuclei.

Solution to Problem 2.24

(a) From Eqs. (2.233), the Bohr radius for the muon is

$$a_0^{(\mu)} = \frac{\hbar^2}{m_\mu e^2} = \frac{m_e}{m_\mu} a_0^{(e)} = \frac{1}{206.8} a_0^{(e)}$$

Since the size of the orbits scales with the Bohr radius, the muon in a $1s$ orbit sits well inside the cloud of atomic electrons.

The nuclear radius is given approximately by[9]

$$R \approx 1.2 A^{1/3} \times 10^{-13} \, \text{cm}$$

From above, the Bohr radius for the muon is

$$a_0^{(\mu)} = \frac{1}{206.8} \times 0.529 \times 10^{-8} \, \text{cm} = 2.56 \times 10^{-11} \, \text{cm}$$

Thus for light nuclei with small $A^{1/3}$, one has

$$R \ll a_0^{(\mu)} \ll a_0^{(e)}$$

(b) The weak interaction through which the muon gets captured is a contact interaction, and the muon capture rate is therefore proportional to the probability the muon sits at the nucleus times the number of protons in the nucleus. For light nuclei this gives a capture rate

$$\Lambda_{\mu\text{-cap}} \propto |\psi_{100}(0)|^2 Z$$
$$\propto Z^4$$

where use has been made of Eq. (2.298).

Problem 2.25 Consider the radial part of the three-dimensional stationary-state Schrödinger equation $R(r) = u(r)/r$ with an attractive

---

[9]See [Walecka (2008)].

exponential potential[10]

$$\frac{d^2u}{dr^2} + \left[\frac{2m_0}{\hbar^2}[E - V(r)] - \frac{l(l+1)}{r^2}\right]u = 0$$

$$V(r) = -V_0 e^{-r/a}$$

(a) Assume s-waves (l=0). Change variables from r to $z \equiv e^{-r/2a}$ and show that Bessel's equation results;

(b) What boundary conditions are to be imposed on $u_0(r)$ as a function of z, and how can these be used to determine the energy levels?

(c) What is the lower limit on V_0 for which a bound state exists?

Solution to Problem 2.25

(a) The bound-state s-wave Schrödinger equation in an attractive exponential potential $V = -V_0 e^{-r/a}$ is

$$\frac{d^2u}{dr^2} + \frac{2m_0}{\hbar^2}\left(V_0 e^{-r/a} - |E|\right)u = 0$$

where we write $E = -|E|$. Now change variables to $z = e^{-r/2a}$, with $dz/dr = -z/2a$, and use the chain rule of differentiation (twice)

$$\frac{du(z)}{dr} = \frac{du(z)}{dz}\frac{dz}{dr} = -\frac{1}{2a}z\frac{du(z)}{dz}$$

$$\frac{d^2u(z)}{dr^2} = \frac{d}{dz}\left[-\frac{1}{2a}z\frac{du(z)}{dz}\right]\frac{dz}{dr}$$

$$= \frac{1}{4a^2}z^2\frac{d^2u(z)}{dz^2} + \frac{1}{4a^2}z\frac{du(z)}{dz}$$

Define

$$\rho \equiv \left[\frac{8m_0 a^2 V_0}{\hbar^2}\right]^{1/2} z \qquad ; \nu \equiv \left[\frac{8m_0 a^2 |E|}{\hbar^2}\right]^{1/2}$$

The Schrödinger equation then becomes

$$\rho^2\frac{d^2u}{d\rho^2} + \rho\frac{du}{d\rho} + (\rho^2 - \nu^2)u = 0$$

This is Bessel's Eq. (2.206).

(b) We have two boundary conditions:

---

[10]See [Schiff (1968)], problem 4.8.

- To be normalizable, the solution must die off as $r \to \infty$, or as $z \to 0$. This gives

$$J_\nu(0) = 0$$

This condition eliminates the solution $J_{-\nu}(z)$ and restricts us to the solution $J_\nu(z)$ in Eq. (2.207);
- The actual radial solution is $u(r)/r$. We have seen that a solution $1/r$ is too singular at the origin, and hence the second boundary condition is $u(r = 0) = 0$. Since $r = 0$ implies $z = 1$, this second boundary condition becomes

$$J_\nu\left(\left[\frac{8m_0a^2V_0}{\hbar^2}\right]^{1/2}\right) = 0 \qquad ; \nu = \left[\frac{8m_0a^2|E|}{\hbar^2}\right]^{1/2}$$

This is now an eigenvalue equation for $|E|$; it is just Eq. (4.38).

(c) For a bound state at zero energy, the condition is

$$J_0\left(\left[\frac{8m_0a^2V_0}{\hbar^2}\right]^{1/2}\right) = 0$$

The smallest value of V_0 that satisfies this condition is given by $J_0(Z_{10}) = 0$, where Z_{10} is the first zero of J_0. Thus the lower limit on V_0 for which a bound state exists is[11]

$$V_0 = \frac{\hbar^2 Z_{10}^2}{8m_0a^2} \qquad ; Z_{10} = 2.4048$$

Problem 2.26[12] Consider the stationary-state Schrödinger equation in an isotropic three-dimensional harmonic-oscillator potential

$$V(r) = -V_0 + \frac{1}{2}kr^2 = -V_0 + \frac{1}{2}kr^2$$

Introduce the length b through

$$\omega_0 = \left(\frac{k}{m_0}\right)^{1/2} \equiv \frac{\hbar}{m_0b^2} \qquad ; q \equiv \frac{r}{b}$$

---

[11]Compare the potential in Eqs. (4.29), used in the example in Sec. 4.1.1.
[12]This problem is longer, but the results are very important, and the steps all parallel those detailed in the text for the hydrogen atom.

(a) Show the solutions can be written in the form

$$\psi_{nlm}(\mathbf{r}) = \frac{u_{nl}(r)}{r} Y_{lm}(\theta, \phi)$$
$$u_{nl}(r) = N_{nl} q^{l+1} e^{-q^2/2} \mathcal{L}_{n-1}^{l+1/2}(q^2)$$

where the *generalized Laguerre polynomials* are defined in terms of the confluent hypergeometric series by[13]

$$\mathcal{L}_p^a(z) \equiv \frac{[\Gamma(a+p+1)]^2}{p!\,\Gamma(a+1)} F(-p|a+1|z)$$
$$= \frac{\Gamma(a+p+1)}{\Gamma(p+1)} \frac{e^z}{z^a} \frac{d^p}{dz^p} \left(z^{a+p} e^{-z}\right)$$

(b) Show the eigenvalue spectrum is given by (see Fig. 2.16)

$$E_{nl} = -V_0 + \hbar\omega_0 \left(N + \frac{3}{2}\right)$$
$$N \equiv 2(n-1) + l \qquad\qquad ; \; n = 1, 2, 3, \cdots, \infty$$
$$l = 0, 1, 2, \cdots, \infty$$

(c) Show the normalization constant is given by

$$N_{nl}^2 = \frac{2(n-1)!}{b[\Gamma(n+l+1/2)]^3}$$

These wave functions form the basis for many nuclear physics calculations.[14]

Solution to Problem 2.26

The discussion here closely parallels that in section 2.4.1 of the text. We look for a solution of the form

$$\psi_{nlm} = \frac{u(r)}{r} Y_{lm}(\theta, \phi)$$

---

[13]Note the *generalized* Laguerre polynomials reduce to the *associated* Laguerre polynomials in the case $a = $ integer.

[14]See [Walecka (2004)].

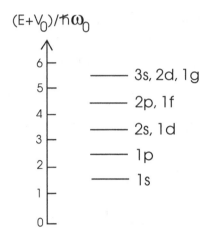

$(E+V_0)/\hbar\omega_0$

6

5 — 3s, 2d, 1g

4 — 2p, 1f

3 — 2s, 1d

2 — 1p

1 — 1s

0

Fig. 2.16   First few levels in the spectrum of the three-dimensional oscillator. This is in the nuclear physics notation (n, l) where these quantum numbers are defined in Prob. 2.26(b). Note that n corresponds to the number of nodes in the radial wave function, excluding the origin and including the point at infinity. (Compare Fig. 2.11 in the text.)

From Eq. (2.201), $u(r)$ satisfies the equation

$$\frac{d^2u}{dr^2} + \left[\frac{2m_0}{\hbar^2}[E - V(r)] - \frac{l(l+1)}{r^2} \right] u = 0$$

$$V(r) = -V_0 + \frac{1}{2}m_0\omega_0^2 r^2$$

Introduce the length b, and dimensionless energy ϵ, through

$$\omega_0 \equiv \frac{\hbar}{m_0 b^2} \qquad ; \ \epsilon \equiv \frac{E + V_0}{\hbar\omega_0/2}$$

With $q \equiv r/b$, the radial equation then takes the form

$$\frac{d^2u}{dq^2} + (\epsilon - q^2)u - \frac{l(l+1)}{q^2}u = 0 \qquad ; \ q \equiv \frac{r}{b}$$

We first take out the acceptable large-q behavior with the introduction of the new function $f(q)$

$$u(q) \equiv f(q)e^{-q^2/2}$$
$$u' = f'e^{-q^2/2} - qfe^{-q^2/2}$$
$$u'' = f''e^{-q^2/2} - 2qf'e^{-q^2/2} - fe^{-q^2/2} + q^2fe^{-q^2/2}$$

With the cancellation of the factor $e^{-q^2/2}$, the radial equation becomes

$$\frac{d^2 f}{dq^2} - 2q\frac{df}{dq} + (\epsilon - 1)f - \frac{l(l+1)}{q^2}f = 0$$

Now take out the small-q behavior with the introduction of

$$f(q) \equiv q^{l+1}g(q)$$
$$f' = q^{l+1}g' + (l+1)q^l g$$
$$f'' = q^{l+1}g'' + 2(l+1)q^l g' + l(l+1)q^{l-1}g$$

With the cancellation of the factor q^l, the radial equation takes the form

$$q\frac{d^2 g}{dq^2} + 2(l+1-q^2)\frac{dg}{dq} + q(\epsilon - 2l - 3)g = 0$$

Finally, introduce

$$g(q) \equiv \frac{1}{4}F(z) \qquad ; z \equiv q^2$$
$$g' = \frac{q}{2}F'$$
$$g'' = zF'' + \frac{1}{2}F'$$

With a cancellation of a factor of q, the radial equation takes the form

$$z\frac{d^2 F}{dz^2} + \left(l + \frac{3}{2} - z\right)\frac{dF}{dz} + \frac{1}{4}(\epsilon - 2l - 3)F = 0$$

This is the confluent hypergeometric Eq. (2.248), and the (unnormalized) solution then takes the form

$$\psi = q^l e^{-q^2/2} F(a|b|z)$$
$$a = -\frac{1}{4}(\epsilon - 2l - 3) \qquad ; b = l + \frac{3}{2} \qquad ; z = q^2$$

The analysis now proceeds exactly as in the text:

- The irregular solution behaves at the origin as $G \sim z^{1-b} = z^{-l-1/2}$, and

$$q^l G \sim q^l (q^2)^{-l-1/2} \sim \frac{1}{q^{l+1}}$$

  This is unacceptable, exactly as in Eq. (2.258); hence, the irregular solution must be discarded;

- Unless the resulting series for $F(a|b|z)$ in Eq. (2.255) terminates, one will find an admixture of the second asymptotic solution $e^{+q^2/2}$, and this is not normalizable. The condition that the series terminate is that a be a negative integer, or zero,

$$-\frac{1}{4}(\epsilon - 2l - 3) = -(n-1) \qquad ; n = 1, 2, 3, \cdots$$

$$\epsilon = 4(n-1) + 2l + 3$$

This implies

$$E + V_0 = \hbar\omega_0 \left[2(n-1) + l + \frac{3}{2}\right]$$

$$\equiv \hbar\omega_0 \left(N + \frac{3}{2}\right) \qquad ; N = 0, 1, 2, 3, \cdots$$

The generalized Laguerre polynomials are defined by

$$\mathcal{L}_p^a(z) \equiv \frac{\Gamma(a+p+1)}{\Gamma(p+1)} \frac{e^z}{z^a} \frac{d^p}{dz^p}\left(z^{a+p}e^{-z}\right)$$

Here p is an integer, and the final function of z is a polynomial of degree p. It can be evaluated with the aid of Leibnitz's formula in Eq. (2.271) as follows

$$\frac{e^z}{z^a} \frac{d^p}{dz^p}\left(z^{a+p}e^{-z}\right) = (a+p)(a+p-1)\cdots(a+1)$$

$$-\binom{p}{1}(a+p)(a+p-1)\cdots(a+2)z + \cdots + (-1)^p z^p$$

The first term comes from putting all the derivatives on the factor z^{a+p}, and the last from putting all the derivatives on e^{-z}. This series is re-written as

$$\frac{e^z}{z^a} \frac{d^p}{dz^p}\left(z^{a+p}e^{-z}\right) = \frac{\Gamma(a+p+1)}{\Gamma(a+1)}\left[1 + \frac{(-p)}{(a+1)}z\right.$$

$$+\frac{(-p)(-p+1)}{(a+1)(a+2)}\frac{z^2}{2!} + \cdots + \left.\frac{(-p)(-p+1)\cdots 1}{(a+1)(a+2)\cdots(a+p)}\frac{z^p}{p!}\right]$$

$$= \frac{\Gamma(a+p+1)}{\Gamma(a+1)}F(-p|a+1|z)$$

The last line identifies the confluent hypergeometric series in Eq. (2.255). Hence, the generalized Laguerre polynomials are expressed in terms of the confluent hypergeometric series by

$$\mathcal{L}_p^a(z) \equiv \frac{\Gamma(a+p+1)}{\Gamma(p+1)} \frac{e^z}{z^a} \frac{d^p}{dz^p} \left(z^{a+p} e^{-z}\right)$$

$$= \frac{[\Gamma(a+p+1)]^2}{p!\,\Gamma(a+1)} F(-p|a+1|z)$$

With the identification of $p = n-1$ and $a = l+1/2$, the previous harmonic oscillator solutions can then be written as

$$\psi_{nlm}(\mathbf{r}) = \frac{u_{nl}(r)}{r} Y_{lm}(\theta, \phi)$$

$$u_{nl}(r) = N_{nl} q^{l+1} e^{-q^2/2} \mathcal{L}_{n-1}^{l+1/2}(q^2)$$

where N_{nl} is now the normalization constant, taken to be real and positive. The normalization condition is

$$b N_{nl}^2 \int_0^\infty dq\, (q^2)^{l+1} e^{-q^2} \left[\mathcal{L}_{n-1}^{l+1/2}(q^2)\right]^2 = 1$$

Insertion of the above representation of $\mathcal{L}_{n-1}^{l+1/2}(q^2)$ gives

$$\frac{b N_{nl}^2}{2} \left[\frac{\Gamma(n+l+1/2)}{(n-1)!}\right]^2 \mathcal{I} = 1$$

where the required integral is, with $z = q^2$ and $p = n-1$,

$$\mathcal{I} \equiv \int_0^\infty dz\, z^{l+1/2} e^{-z} \left[\frac{e^z}{z^{l+1/2}} \frac{d^p}{dz^p} \left(z^{p+l+1/2} e^{-z}\right)\right]$$

$$\times \left[\frac{e^z}{z^{l+1/2}} \frac{d^p}{dz^p} \left(z^{p+l+1/2} e^{-z}\right)\right]$$

$$= \int_0^\infty dz\, \frac{e^z}{z^{l+1/2}} \left[\frac{d^p}{dz^p} \left(z^{p+l+1/2} e^{-z}\right)\right] \left[\frac{d^p}{dz^p} \left(z^{p+l+1/2} e^{-z}\right)\right]$$

Now carry out p partial integrations on z [15]

$$\mathcal{I} = (-1)^p \int_0^\infty dz\, z^{p+l+1/2} e^{-z} \frac{d^p}{dz^p} \left[\frac{e^z}{z^{l+1/2}} \frac{d^p}{dz^p} \left(z^{p+l+1/2} e^{-z}\right)\right]$$

The quantity in square brackets is a polynomial of degree p

$$\left[\frac{e^z}{z^{l+1/2}} \frac{d^p}{dz^p} \left(z^{p+l+1/2} e^{-z}\right)\right] = a_p z^p + a_{p-1} z^{p-1} + \cdots + a_0$$

---

[15] We leave it for readers to convince themselves that the boundary contributions vanish.

Only the first term contributes when the next set of derivatives is taken, and a_p comes from placing all the present derivatives on the exponential

$$a_p = (-1)^p$$

The result is

$$\mathcal{I} = p! \int_0^\infty dz\, z^{p+l+1/2}\, e^{-z} = p!\, \Gamma(p + l + 3/2)$$

The normalization condition then becomes, with $p = n - 1$,

$$N_{nl}^2 = \frac{2(n-1)!}{b[\Gamma(n+l+1/2)]^3}$$

Problem 2.27 (a) The isotropic three-dimensional harmonic-oscillator potential in Prob. 2.26 can be re-written using $\mathbf{r}^2 = x^2 + y^2 + z^2$. Show the problem separates into three one-dimensional oscillators with an eigenvalue spectrum

$$E_{n_x n_y n_z} = -V_0 + \hbar\omega_0 \left(N + \frac{3}{2} \right)$$

$$N \equiv n_x + n_y + n_z \qquad\qquad ; \; (n_x, n_y, n_z) = 0, 1, 2, \cdots, \infty$$

(b) Show the degeneracy of the levels with $N = 0, \cdots, 4$ is exactly that obtained in Prob. 2.26(b).

Solution to Problem 2.27

(a) The time-independent Schrödinger equation in cartesian coordinates here is

$$\left[-\frac{\hbar^2}{2m_0} \left(\frac{\partial^2}{\partial x^2} + \frac{\partial^2}{\partial y^2} + \frac{\partial^2}{\partial z^2} \right) + \frac{1}{2} m_0 \omega_0^2 \left(x^2 + y^2 + z^2 \right) \right] \psi = (E + V_0) \psi$$

The separated solution to this equation takes the form of a product of simple harmonic oscillator solutions

$$\psi(x, y, z) = \psi_{n_x}(x) \psi_{n_y}(y) \psi_{n_z}(z)$$

The corresponding eigenvalue is simply the sum of the eigenvalues

$$E_{n_x n_y n_z} + V_0 = \hbar\omega_0 \left(n_x + n_y + n_z + \frac{3}{2} \right) \qquad ; \; (n_x, n_y, n_z) = 0, 1, 2, \cdots, \infty$$

(b) The degeneracy in the three-dimensional simple harmonic oscillator in radial coordinates is computed from the result in Prob. 2.26

$$N = 2(n-1) + l \quad ; \ n = 1, 2, 3, \cdots \quad\quad ; \ l = 0, 1, 2, \cdots$$
$$; \ m = 0, \pm 1, \cdots, \pm l$$

In cartesian coordinates, the degeneracy is computed from

$$N = n_x + n_y + n_z \quad ; \ (n_x, n_y, n_z) = 0, 1, 2, \cdots$$

The degeneracies for the first 5 levels are shown in Table 2.1. They are the same in either coordinate system.

Table 2.1 Degeneracies in the 3-D oscillator in radial or cartesian coordinates.

N	radial coord.	cartesian coord.
0	1	1
1	3	3
2	6	6
3	10	10
4	15	15

Problem 2.28 Compute and plot the radial wave functions for the first four states in Fig. 2.16 as a function of r/b. Compare with the results in Fig. 2.12 in the text.

Solution to Problem 2.28

The solutions for the isotropic harmonic oscillator were found in Prob. 2.26. They are

$$\psi_{nlm}(\mathbf{r}) = \frac{u_{nl}(r)}{r} Y_{lm}(\theta, \phi)$$
$$u_{nl}(r) = N_{nl} q^{l+1} e^{-q^2/2} \mathcal{L}_{n-1}^{l+1/2}(q^2)$$

Here $\mathcal{L}_{n-1}^{l+1/2}(q^2)$ are the generalized Laguerre polynomials defined by

$$\mathcal{L}_p^a(z) \equiv \frac{\Gamma(a+p+1)}{\Gamma(p+1)} \frac{e^z}{z^a} \frac{d^p}{dz^p} \left(z^{a+p} e^{-z} \right)$$
$$= \frac{[\Gamma(a+p+1)]^2}{p!\,\Gamma(a+1)} F(-p|a+1|z)$$

The normalization constant is given by

$$N_{nl}^2 = \frac{2(n-1)!}{b[\Gamma(n+l+1/2)]^3}$$

We use these formulas to obtain the radial wave functions of the first four states in terms of $x \equiv r/b$

$$b^{3/2}\frac{u_{10}(r)}{r} = \frac{2}{\pi^{1/4}}e^{-x^2/2} \qquad\qquad ; \, x = \frac{r}{b}$$

$$b^{3/2}\frac{u_{11}(r)}{r} = \frac{2}{\pi^{1/4}}\sqrt{\frac{2}{3}}\, xe^{-x^2/2}$$

$$b^{3/2}\frac{u_{12}(r)}{r} = \frac{4}{\pi^{1/4}}\sqrt{\frac{1}{15}}\, x^2 e^{-x^2/2}$$

$$b^{3/2}\frac{u_{20}(r)}{r} = \frac{\sqrt{6}}{\pi^{1/4}}\left(1 - \frac{2x^2}{3}\right)e^{-x^2/2}$$

These wave functions are plotted in Fig. 2.17. They can be compared with the radial wave functions of a particle in a spherical box plotted in Fig. 2.12 of the text. To allow a better comparison, in Fig. 2.18 we have plotted the wave functions $a^{3/2}N_{nl}j_l(X_{nl}r/a)$ as a function of r/a.[16]

In the case of a particle in a spherical box in Fig. 2.18, one observes that all the wave functions vanish at the surface of the box as a result of the infinite potential barrier. In the case of the isotropic harmonic oscillator in Fig. 2.17, the wave functions are not confined to a finite region of space, although they decay exponentially at large distances. The effect of the confinement is greater in the case of the 2s-state.

Problem 2.29 (a) In two dimensions, in polar coordinates, the laplacian is[17]

$$\nabla^2 = \frac{1}{r}\frac{\partial}{\partial r}r\frac{\partial}{\partial r} + \frac{1}{r^2}\frac{\partial^2}{\partial \phi^2}$$

Show that for a particle in a constant potential $-V_0$ in two dimensions, the separated solutions to the time-independent Schrödinger equation take the form

$$\psi(r,\phi) = R(r)e^{\pm im\phi} \qquad ; \, m = 0,1,2,\cdots$$

$$z^2\frac{d^2R}{dz^2} + z\frac{dR}{dz} + (z^2 - m^2)R = 0 \qquad ; \, z \equiv \kappa r$$

---

[16]Notice that in the text only the spherical Bessel functions are plotted.
[17]See Vol. I.

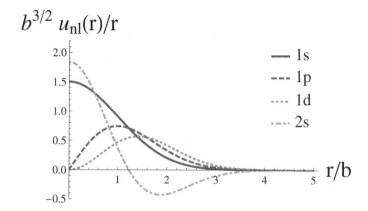

Fig. 2.17 Radial wave functions for the $1s$-, $1p$-, $1d$-, and $2s$-states as a function of r/b for the three-dimensional oscillator.

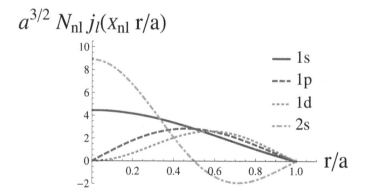

Fig. 2.18 Radial wave functions for the $1s$-, $1p$-, $1d$-, and $2s$-states as a function of r/a for the spherical cavity.

where $\kappa^2 \equiv (2m_0/\hbar^2)(E + V_0)$;

(b) A fundamental system of solutions to the above radial equation is given by the cylindrical Bessel function $J_m(z)$ of Eq. (2.207) and the cylindrical Neumann function $N_m(z)$.[18] The Neumann function is singular at the origin and has a wronskian $W[J_m(z), N_m(z)] = 2/\pi z$. Construct an argument that eliminates the singular solution.

---

[18]See [Fetter and Walecka (2003a)].

Solution to Problem 2.29

(a) The stationary-state Schrödinger equation reads

$$(\nabla^2 + \kappa^2)\psi = 0$$

$$\kappa^2 \equiv \left(\frac{2m_0}{\hbar^2}\right)(E + V_0)$$

with the laplacian in two dimensions given by

$$\nabla^2 = \frac{1}{r}\frac{\partial}{\partial r}r\frac{\partial}{\partial r} + \frac{1}{r^2}\frac{\partial^2}{\partial \phi^2}$$

Look for a separated solution

$$\psi = R(r)\Phi(\phi)$$

where, as in Eqs. (2.146) and (2.150),

$$\Phi_{\pm m}(\phi) = \frac{1}{\sqrt{2\pi}}e^{\pm im\phi} \qquad ; m = 0, 1, 2, \cdots$$

Substitution into the Schrödinger equation, cancellation of Φ, and multiplication by r^2 then gives

$$r^2\frac{d^2 R}{dr^2} + r\frac{dR}{dr} + (\kappa^2 r^2 - m^2)R = 0$$

This is Bessel's Eq. (2.206)

$$z^2\frac{d^2 R}{dz^2} + z\frac{dR}{dz} + (z^2 - \nu^2)R = 0 \qquad ; z \equiv \kappa r \qquad ; \nu \equiv m$$

(b) A fundamental system of solutions to this equation for integer m consists of the cylindrical Bessel and Neumann functions $\{J_m(z), N_m(z)\}$, with wronskian $W[J_m(z), N_m(z)] = 2/\pi z$.[19] The Neumann function is singular at the origin. Suppose $J_m(z)$ is an acceptable solution, and consider the superposed solution $\psi = [aJ_m(z) + bN_m(z)]\Phi_m(\phi)$. The net probability flux out of a very small circle surrounding the origin is, with $d\mathbf{A} = \mathbf{e}_r R d\phi$,

$$\oint_C \mathbf{S} \cdot d\mathbf{A} = \frac{\hbar\kappa R}{2im_0}\left\{[aJ_m(z) + bN_m(z)]^\star\left[a\frac{dJ_m(z)}{dz} + b\frac{dN_m(z)}{dz}\right]\right.$$

$$\left. - \left[a\frac{dJ_m(z)}{dz} + b\frac{dN_m(z)}{dz}\right]^\star[aJ_m(z) + bN_m(z)]\right\}_{z=\kappa R}$$

$$= \frac{\hbar\,\mathrm{Im}\,(a^\star b)}{m_0}\{zW[J_m(z), N_m(z)]\}_{z=\kappa R}$$

---

[19]See [Fetter and Walecka (2003a); Morse and Feshbach (1953)]; see also Prob. 2.19.

Therefore

$$\oint_C \mathbf{S} \cdot d\mathbf{A} = \frac{2\hbar \operatorname{Im}(a^* b)}{\pi m_0}$$

This expression is finite as $R \to 0$. The origin thus acts as a point source of probability, and hence the singular solution $N_m(z)$ must be discarded.

Problem 2.30 (a) Use the results in Prob. 2.29 to show that in two dimensions the eigenvalues and eigenfunctions for a particle in a circular box of radius a are given by

$$E_{nm} = -V_0 + \frac{\hbar^2}{2m_0 a^2} Z_{nm}^2 \qquad ; m = 0, 1, 2, \cdots, \infty$$
$$; n = 1, 2, \cdots, \infty$$
$$\psi_{n, \pm m} = N_{nm} J_m \left(Z_{nm} \frac{r}{a} \right) e^{\pm im\phi}$$

Here Z_{nm} is the nth zero of the mth cylindrical Bessel function, excluding the origin;

(b) Plot the first four eigenvalues and corresponding radial wave functions $J_m(Z_{nm} r/a)$.

Solution to Problem 2.30

(a) From the results of Prob. 2.29, we see that the wave function of a particle in a circular box of radius a may be written as

$$\psi = R(r)\Phi(\phi)$$

where

$$\Phi_{\pm m}(\phi) = \frac{1}{\sqrt{2\pi}} e^{\pm im\phi} \qquad ; m = 0, 1, 2, \cdots$$

while the radial wave function $R(r)$ obeys Bessel's equation

$$r^2 \frac{d^2 R}{dr^2} + r \frac{dR}{dr} + (\kappa^2 r^2 - m^2) R = 0$$

The general solution to this equation reads

$$R(r) = c_1 J_m(\kappa r) + c_2 N_m(\kappa r)$$

where $\kappa^2 \equiv (2m_0/\hbar^2)(E + V_0)$. Since the Bessel functions of the second kind are singular at $r = 0$, they cannot appear in the solution for a particle in a circular box, and we must choose $c_2 = 0$.[20] The remaining constant c_1 is found by imposing the normalization of the full wave function.

Since the particle feels an infinite potential wall at $r = a$, one must also impose the boundary condition that the wave function vanish at the surface of the circular box, and therefore one may write

$$R(r) = c_1 J_m \left(Z_{nm} \frac{r}{a} \right) \qquad ; \ m = 0, 1, 2, \cdots, \infty$$
$$\qquad ; \ n = 1, 2, 3, \cdots, \infty$$

where Z_{nm} is the nth zero of J_m, excluding the origin.

The complete wave function is therefore

$$\psi_{n,\pm m} = N_{nm} J_m \left(Z_{nm} \frac{r}{a} \right) e^{\pm im\phi}$$

where N_{nm} is now the normalization constant.[21] In particular one has

$$N_{nm}^2 = \left[\int_0^a r \, dr \, J_m^2 \left(Z_{nm} \frac{r}{a} \right) \right]^{-1}$$
$$= \frac{1}{\pi a^2 J_{m+1}^2 (Z_{nm})}$$

The identification $Z_{nm}/a \to \kappa_{nm} = \sqrt{(2m_0/\hbar^2)(E_{nm} + V_0)}$ allows one to obtain the eigenvalues as

$$E_{nm} = -V_0 + \frac{\hbar^2}{2m_0 a^2} Z_{nm}^2$$

(b) In Fig. 2.19 we plot the quantity $2m_0 a^2 (E_{nm} + V_0)/\hbar^2$ involving the first four eigenvalues of a particle trapped in a circular box in two dimensions. Here $\hbar m$ is the angular momentum about the z-axis. Notice that the states with $m \neq 0$ are doubly degenerate, corresponding to the particle moving in either a clockwise or counter-clockwise direction

The corresponding radial wave functions are plotted in Fig. 2.20. It is interesting to compare these wave functions with those for a particle in a spherical box in Fig. 2.18. Again, the angular-momentum barrier moves the higher m-states toward the wall of the container.

---

[20]See Prob. 2.29. These solutions do appear in the wave function for a particle trapped in an annular region, since in this case the origin is not part of the physical domain.

[21]This is the constant c_1 that we had written before; however, we have changed the notation to take into account the fact that the normalization constant is different for each state.

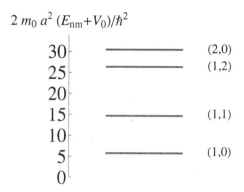

Fig. 2.19 The quantity $2m_0a^2(E_{nm}+V_0)/\hbar^2$ for the four lowest eigenvalues of a particle trapped in a circular box in two dimensions. The levels are identified by (n, m).

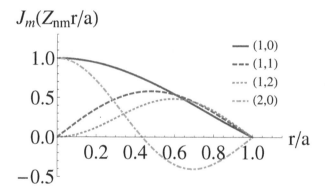

Fig. 2.20 Radial wave functions for the states corresponding to the eigenvalues in Fig. 2.19.

Problem 2.31 Solve the eigenvalue Eqs. (2.336) numerically for a few values of λ, and plot the curves in the band structure and extended band structure of the Kronig-Penney model in Figs. 2.21 and 2.22 of the text.

Solution to Problem 2.31

We consider the eigenvalue equation Eq. (2.336)

$$\cos qa = F(\xi)$$
$$F(\xi) = \cos\xi + \frac{\lambda}{2}\frac{\sin\xi}{\xi}$$

where

$$qa = \frac{2\pi p}{N} \qquad ; p = 0, \pm 1, \cdots, \pm \frac{N}{2}$$

Clearly, for a given value of qa, the equation admits multiple solutions (see Fig. 2.20 in the text); call these solutions ξ_n. Notice that for $\xi \to \infty$, one has $\sin \xi / \xi \to 0$, and therefore $\xi_n \to qa + 2n\pi$. Moreover, the solutions corresponding to $p = \pm N/2$, that is $qa = \pm \pi$, are simply $\xi_n = \pm \pi + 2n\pi$.

Table 2.2 First four numerical solutions of the eigenvalue Eq. (2.336)
for the Kronig-Penney model, for
$\lambda = 10$ and $qa = 1$.

ξ_1	ξ_2	ξ_3	ξ_4
2.447	5.843	7.941	11.914

In Fig. 2.21 we plot the graphical solution to the eigenvalue Eq. (2.336) for the Kronig-Penney model, for $\lambda = 10$ and $qa = 1$. The numerical values of the solutions are displayed in Table 2.2.

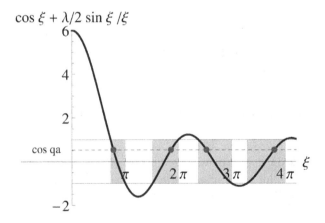

Fig. 2.21   Graphical solution to the eigenvalue Eq. (2.336) for the Kronig-Penney model, for $\lambda = 10$ and $qa = 1$.

In Fig. 2.22 we plot the numerical calculation of the energy bands in the Kronig-Penney model for $\lambda = 10$.

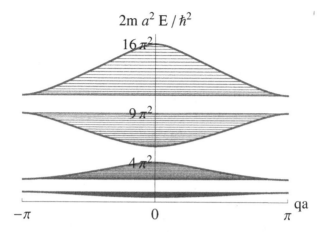

Fig. 2.22 Energy bands in the Kronig-Penney model for $\lambda = 10$.

The numerical calculation of the extended band structure in the Kronig-Penney model for $\lambda = 10$ is plotted in Fig. 2.23.

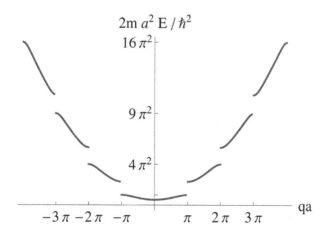

Fig. 2.23 Extended band structure in the Kronig-Penney model for $\lambda = 10$.

Problem 2.32 As a preliminary to the next two problems, this problem reviews and extends the discussion in Vol. I of barrier penetration in one dimension. Assume a repulsive step potential of height V_0 and width l, with $E < V_0$ (see Fig. 2.24).

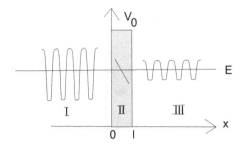

Fig. 2.24 Barrier penetration in one dimension, with $E < V_0$. There are both incident and reflected waves $e^{ikx} + re^{-ikx}$ in region I, and a transmitted wave te^{ikx} in region III.

The Schrödinger equation and solutions in the various regions are given in Eqs. (2.24)–(2.29). With the utilization of scattering states, the solutions in the three regions take the following form

$$\psi_I = e^{ikx} + re^{-ikx} \qquad ; \ \psi_{II} = ae^{-\kappa x} + be^{\kappa x} \qquad ; \ \psi_{III} = te^{ikx}$$

(a) Match the boundary conditions at both sides of the potential, and show that the transmission and reflection amplitudes satisfy the following relations

$$te^{ikl} = e^{-\kappa l} + re^{-\kappa l}e^{-i\xi} \qquad ; \ e^{i\xi} \equiv \frac{\kappa - ik}{\kappa + ik}$$
$$te^{ikl} = e^{\kappa l} + re^{\kappa l}e^{i\xi}$$

(b) Show the transmission amplitude and transmission coefficient are

$$t = e^{-ikl}\left[\frac{e^{i\xi} - e^{-i\xi}}{e^{\kappa l}e^{i\xi} - e^{-\kappa l}e^{-i\xi}}\right]$$
$$T \equiv |t|^2 = \frac{(2k\kappa)^2}{(2k\kappa)^2 + (\kappa^2 + k^2)^2 \sinh^2 \kappa l}$$

(c) Show the reflection amplitude and reflection coefficient are

$$r = -\left[\frac{e^{\kappa l} - e^{-\kappa l}}{e^{\kappa l}e^{i\xi} - e^{-\kappa l}e^{-i\xi}}\right]$$
$$R \equiv |r|^2 = \frac{(\kappa^2 + k^2)^2 \sinh^2 \kappa l}{(2k\kappa)^2 + (\kappa^2 + k^2)^2 \sinh^2 \kappa l}$$

(d) Verify that $R + T = 1$, and interpret this result.

Solution to Problem 2.32

(a) The conditions that the value and slope be continuous between regions I and II are

$$1 + r = a + b$$
$$ik(1 - r) = -\kappa(a - b)$$

Similarly, between regions II and III

$$t^{ikl} = ae^{-\kappa l} + be^{\kappa l}$$
$$ikte^{ikl} = -\kappa(ae^{-\kappa l} - be^{\kappa l})$$

These four equations are re-written as

$$1 + r = a + b$$
$$(-ik/\kappa)(1 - r) = a - b$$
$$t^{ikl} = ae^{-\kappa l} + be^{\kappa l}$$
$$(-ik/\kappa)te^{ikl} = ae^{-\kappa l} - be^{\kappa l}$$

Addition and subtraction of the first two gives

$$2a = (1 - ik/\kappa) + r(1 + ik/\kappa)$$
$$2b = (1 + ik/\kappa) + r(1 - ik/\kappa)$$

Addition and subtraction of the last two, and the use of these expressions, gives two relations between t and r

$$t(1 - ik/\kappa)e^{ikl} = e^{-\kappa l}[(1 - ik/\kappa) + r(1 + ik/\kappa)]$$
$$t(1 + ik/\kappa)e^{ikl} = e^{\kappa l}[(1 + ik/\kappa) + r(1 - ik/\kappa)]$$

Now define the real phase ξ through

$$e^{i\xi} \equiv \frac{\kappa - ik}{\kappa + ik}$$

The previous two relations then become

$$te^{ikl} = e^{-\kappa l} + re^{-\kappa l}e^{-i\xi}$$
$$te^{ikl} = e^{\kappa l} + re^{\kappa l}e^{i\xi}$$

(b) One can eliminate r from the above equations by first writing

$$te^{ikl}e^{\kappa l}e^{i\xi} = e^{i\xi} + r$$
$$te^{ikl}e^{-\kappa l}e^{-i\xi} = e^{-i\xi} + r$$

and then subtracting these two equations. This gives

$$t = e^{-ikl} \left[\frac{e^{i\xi} - e^{-i\xi}}{e^{\kappa l} e^{i\xi} - e^{-\kappa l} e^{-i\xi}} \right]$$

To compute the absolute square of this quantity, it is first convenient to rewrite it as[22]

$$t = e^{-ikl} \left[\frac{(\kappa - ik)/(\kappa + ik) - (\kappa + ik)/(\kappa - ik)}{e^{\kappa l}(\kappa - ik)/(\kappa + ik) - e^{-\kappa l}(\kappa + ik)/(\kappa - ik)} \right]$$

$$= e^{-ikl} \left[\frac{-4ik\kappa}{e^{\kappa l}(\kappa^2 - k^2 - 2ik\kappa) - e^{-\kappa l}(\kappa^2 - k^2 + 2ik\kappa)} \right]$$

$$= e^{-ikl} \left[\frac{2k\kappa}{2k\kappa \cosh \kappa l + i(\kappa^2 - k^2)\sinh \kappa l} \right]$$

Thus the transmission coefficient for the barrier is given by

$$T = |t|^2 = \frac{(2k\kappa)^2}{(2k\kappa)^2 + (\kappa^2 + k^2)^2 \sinh^2 \kappa l}$$

(c) A subtraction of the last two equations in part (a) leads to a result for r

$$r = -\frac{e^{\kappa l} - e^{-\kappa l}}{e^{\kappa l} e^{i\xi} - e^{-\kappa l} e^{-i\xi}}$$

As above, this can be re-written as

$$r = -\frac{2\sinh \kappa l}{e^{\kappa l}(\kappa - ik)/(\kappa + ik) - e^{-\kappa l}(\kappa + ik)/(\kappa - ik)}$$

$$= -\frac{(\kappa^2 + k^2)\sinh \kappa l}{(\kappa^2 - k^2)\sinh \kappa l - 2ik\kappa \cosh \kappa l}$$

Hence the reflection coefficient for the barrier is

$$R = |r|^2 = \frac{(\kappa^2 + k^2)^2 \sinh^2 \kappa l}{(2k\kappa)^2 + (\kappa^2 + k^2)^2 \sinh^2 \kappa l}$$

(d) It is evident that

$$T + R = 1$$

This is the statement of conservation of probability; in this stationary state, the particle can either be transmitted or reflected by the barrier.

---

[22] Use $\sinh x = (e^x - e^{-x})/2$, $\cosh x = (e^x + e^{-x})/2$, and $\cosh^2 x - \sinh^2 x = 1$.

Problem 2.33 To obtain the boundary condition on the wave function at a delta-function potential, take the following limit of the results in Prob. 2.32

$$V_0 \to \infty \qquad ; \, l \to 0 \qquad ; \int V(x)dx = V_0 l = \text{constant}$$

(a) Show $t = 1 + r$;
(b) Hence conclude that the wave function must be continuous across the potential, $\psi_{III}(0) = \psi_I(0)$.

Solution to Problem 2.33

For the given limit of a delta-function potential, where $v_0 \to \infty$, $l \to 0$, while $v_0 l$ is held fixed and finite[23]

$$\kappa \to \sqrt{v_0} \to \infty$$
$$\kappa l \to \sqrt{v_0 l^2} \to 0$$

Thus in this limit

$$e^{i\xi} \to 1$$
$$e^{ikl} \to 1$$

Therefore both equations at the end of the solution to Prob. 2.32(a) reduce to

$$t = 1 + r$$

This implies

$$\psi_{III}(0) = \psi_I(0)$$

Hence the wave function is continuous across the delta-function potential.

Problem 2.34 Use the results in Probs. 2.32–2.33 to show that $T \to 0$ as $\lambda \to \infty$ for the delta-function potential in Eq. (2.315), and hence demonstrate that there is no barrier penetration in the tight-binding limit of the Kronig-Penney model.

---

[23] Here $v_0 = 2mV_0/\hbar^2$.

Solution to Problem 2.34

In the limit of a delta-function potential, the results in Prob. 2.32–2.33 imply that the transmission amplitude goes to

$$t \to \frac{1}{1 + i\kappa^2 l/2k}$$

Consider the specific delta-function potential used in the text, with $v_0 l = 2mV_0 l/\hbar^2 = \lambda/a$. The above expression becomes

$$t \to \frac{1}{1 + i\lambda/2ka}$$

This vanishes as the strength of the delta-function potential $\lambda \to \infty$

$$t \to 0 \qquad ; \lambda \to \infty$$

We conclude that there is no barrier penetration in the tight-binding limit of the Kronig-Penney model.

Problem 2.35 A particle of mass m is confined by rigid walls to the interior of a rectangular box with dimensions $0 \le x \le a$, $0 \le y \le a$, and $0 \le z \le l$ (see Fig. 2.25).

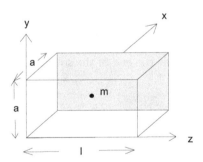

Fig. 2.25   Particle of mass m in a rectangular box with dimensions $0 \le x \le a$, $0 \le y \le a$, and $0 \le z \le l$.

(a) Construct the general solution to the time-dependent Schrödinger equation;

(b) Suppose the particle is injected into the box at time $t = 0$ in such a manner that its wave function at that time is

$$\Psi(\mathbf{x}, 0) = \left(\frac{8}{a^2 l} \right)^{1/2} \sin\frac{\pi x}{a} \sin\frac{\pi y}{a} \left[\sin\frac{\pi z}{l} + \sin\frac{2\pi z}{l} \right]$$

Plot the initial density distribution, and determine the wave function at all subsequent times;

(c) Compute the integrated probability flux through the plane $z = l/2$ as a function of time. Then use the continuity equation to compute the probability of finding the particle in the right-hand half of the box as a function of time. Plot the latter quantity. After what time (or times) will it be maximized?

Solution to Problem 2.35

(a) The general solution to the time-dependent Schrödinger equation can be expressed in terms of the solutions of the time-independent Schrödinger equation as

$$\Psi(x, y, z, t) = \sum_{n_x, n_y, n_z} c_{n_x n_y n_z} e^{-iE_{n_x n_y n_z} t/\hbar} \psi_{n_x n_y n_z}(x, y, z)$$

where $n_i = 1, 2, \cdots$, and $c_{n_x n_y n_z}$ are complex coefficients. Here $E_{n_x n_y n_z}$ and $\psi_{n_x n_y n_z}(x, y, z)$ are the eigenvalues and eigenfunctions of the free hamiltonian of the particle in the rectangular box; they read

$$E_{n_x n_y n_z} = \frac{\hbar^2 \pi^2}{2m} \left(\frac{n_x^2}{a^2} + \frac{n_y^2}{a^2} + \frac{n_z^2}{l^2} \right)$$

$$\psi_{n_x n_y n_z}(x, y, z) = \left(\frac{8}{a^2 l} \right)^{1/2} \sin \frac{n_x \pi x}{a} \sin \frac{n_y \pi y}{a} \sin \frac{n_z \pi z}{l}$$

Observe that these eigenfunctions are properly normalized

$$\int_0^a dx \int_0^a dy \int_0^l dz \left| \psi_{n_x, n_y, n_z}(x, y, z) \right|^2 = 1$$

The normalization of $\Psi(x, y, z, t)$ implies

$$\sum_{n_x, n_y, n_z} \left| c_{n_x, n_y, n_z} \right|^2 = 1$$

(b) First of all, it is useful to express the given initial wave function in terms of the eigenfunctions of the hamiltonian

$$\Psi(\mathbf{x}, 0) = \psi_{111}(x, y, z) + \psi_{112}(x, y, z)$$

Note that this wave function is not normalized, since

$$\int_0^a dx \int_0^a dy \int_0^l dz |\Psi(\mathbf{x}, 0)|^2 = 2$$

In order to compute physical quantities, it is more convenient to work from the outset with a normalized $\Psi(\mathbf{x}, 0)$. This is readily accomplished through the substitution

$$\Psi(\mathbf{x}, 0) \to \frac{1}{\sqrt{2}} \left[\psi_{111}(x, y, z) + \psi_{112}(x, y, z) \right] \qquad ; \text{ normalized}$$

We then easily read off the coefficients $c_{111} = c_{112} = 1/\sqrt{2}$; all the remaining coefficients vanish. The normalized wave function at all subsequent times is therefore

$$\Psi(\mathbf{x}, t) = \frac{1}{\sqrt{2}} \left[e^{-iE_{111}t/\hbar} \psi_{111}(x, y, z) + e^{-iE_{112}t/\hbar} \psi_{112}(x, y, z) \right]$$

In Fig. 2.26 we plot the initial density distribution $a^2 l |\Psi(x, y, z, 0)|^2$ at $x = y = a/2$ as a function of z. The probability is peaked in the l.h.s. of the box, as viewed in Fig. 2.25.

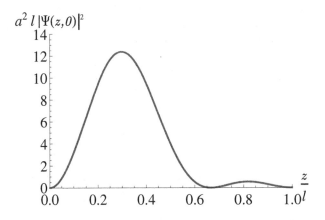

Fig. 2.26   Initial density distribution at $x = y = a/2$ as a function of z.

We use the above wave function to calculate the probability current, given in Eq. (2.7) of the text

$$\mathbf{S} = \frac{\hbar}{2im} \left\{ \Psi^\star(\mathbf{x}, t) \boldsymbol{\nabla} \Psi(\mathbf{x}, t) - \left[\boldsymbol{\nabla} \Psi^\star(\mathbf{x}, t) \right] \Psi(\mathbf{x}, t) \right\}$$

In particular, we are interested in the z-component of this vector

$$S_z(x, y, z, t) = \frac{8\pi\hbar}{a^2 l^2 m} \sin^2\left(\frac{\pi x}{a}\right) \sin^2\left(\frac{\pi y}{a}\right) \sin^3\left(\frac{\pi z}{l}\right) \sin\left(\frac{3\pi^2 \hbar t}{2l^2 m}\right)$$

Notice that **S** vanishes on the wall of the box, because of the boundary conditions obeyed by the wave function (the particle cannot escape from the box). The total current flowing through the surface perpendicular to the z-axis, at $z = l/2$, is

$$\int_\Sigma \mathbf{S} \cdot \mathbf{n} \, d\Sigma = \int_0^a dx \int_0^a dy \, S_z(x, y, l/2, t)$$
$$= \frac{2\pi\hbar}{l^2 m} \sin\left(\frac{3\pi^2 \hbar t}{2l^2 m}\right)$$

Here Σ is the (x, y) surface inside the box at $z = l/2$ and **n** is a unit vector normal to that surface pointing along the positive z-axis.

In fact, the above integral can be extended to include all the walls of the box in the l.h.s. since the current vanishes on those walls. One can then use the continuity equation, invoking Gauss's law as done in Eq. (2.11) of the book, to obtain the rate of change of the total probability in the l.h.s. of the box as

$$-\frac{d}{dt} \int_V \rho \, d^3x = \int_\Sigma \mathbf{S} \cdot \mathbf{n} \, d\Sigma = \frac{2\pi\hbar}{l^2 m} \sin\left(\frac{3\pi^2 \hbar t}{2l^2 m}\right)$$

After integrating this rate of change over t, we obtain the probability of finding the particle in the r.h.s of the box as a function of time

$$P_R(t) = P_0 - \frac{4}{3\pi} \cos\left(\frac{3\pi^2 \hbar t}{2l^2 m}\right)$$

where P_0 is a constant of integration. This constant can be determined in the following way. At $t = 0$ the probability has a minimum, while it reaches a maximum at $t = 2l^2 m/3\pi\hbar$. In the absence of external forces, one must find that at the time $t_0 = l^2 m/3\pi\hbar$ it is equally probable to find the particle in the left or in the right half of the box; we therefore conclude that $P_0 = 1/2$ and[24]

$$P_R(t) = \frac{1}{2} - \frac{4}{3\pi} \cos\left(\frac{3\pi^2 \hbar t}{2l^2 m}\right)$$

This probability is plotted as a function of time in Fig. 2.27. Notice that the probability sloshes back and forth in the box with a period

$$\tau = \frac{2\pi\hbar}{(E_{112} - E_{111})} = \frac{4l^2 m}{3\pi\hbar}$$

---

[24]This result is obtained in a more direct, but less informative fashion from $P_R(t) = \int_0^a dx \int_0^a dy \int_{l/2}^l dz \, |\Psi(\mathbf{x}, t)|^2$.

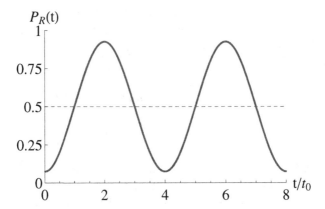

Fig. 2.27   Probability of finding the particle in the r.h.s. of the box as a function of time. The abscissa is t/t_0 where $t_0 = l^2 m/3\pi\hbar = \tau/4$.

Chapter 3

Formal Developments

Problem 3.1 Consider a two-dimensional problem with coefficient matrix

$$\begin{bmatrix} (H_{11} - E) & H_{12} \\ H_{12}^{\star} & (H_{22} - E) \end{bmatrix}$$

(a) Determine the eigenvalues $E^{(s)}$ and eigenvectors $\underline{c}^{(s)}$, for $s = 1, 2$;

(b) Set $H_{11} = H_{22}$ and let $H_{12} \to 0$ so that the eigenvalues become *degenerate*. Show that information is lost, and *any* new orthonormal pair of eigenvectors then provides a solution.

Solution to Problem 3.1

(a) Consider the general case with $H_{11} > H_{22}$ and $H_{12} \neq 0$. The eigenvalues are determined by setting the determinant equal to zero

$$\begin{vmatrix} (H_{11} - E) & H_{12} \\ H_{12}^{\star} & (H_{22} - E) \end{vmatrix} = E^2 - (H_{11} + H_{22})E + H_{11}H_{22} - |H_{12}|^2 = 0$$

The solutions to this quadratic equation are

$$E_{\pm} = \frac{1}{2} \left\{ (H_{11} + H_{22}) \pm \left[(H_{11} - H_{22})^2 + 4|H_{12}|^2 \right]^{1/2} \right\}$$

The limiting cases of this expression as $H_{12} \to 0$ are then

$$E_+ \to H_{11} \qquad ; H_{12} \to 0$$
$$E_- \to H_{22}$$

The corresponding (un-normalized) eigenvectors are obtained from the remaining linear equations

$$\frac{c_2^{(+)}}{c_1^{(+)}} = \frac{H_{12}^\star}{E_+ - H_{22}} \qquad ; \qquad \frac{c_1^{(-)}}{c_2^{(-)}} = \frac{H_{12}}{E_- - H_{11}}$$

The limiting cases of these expression as $H_{12} \to 0$, with $H_{11} \neq H_{22}$, are

$$c_2^{(+)} \to 0 \qquad ; \quad H_{12} \to 0$$
$$c_1^{(-)} \to 0$$

The eigenstate with E_+ is just the first state, and with E_-, the second.

(b) Now suppose $H_{12} = 0$, and let $H_{11} \to H_{22} \equiv E_0$. In this case, the eigenvalue equation becomes

$$\begin{vmatrix} (E_0 - E) & 0 \\ 0 & (E_0 - E) \end{vmatrix} = (E - E_0)^2 = 0$$

The eigenvalues are degenerate, with $E = E_0$, and now there are no remaining restrictions on the ratios of the eigenstate coefficients.

Problem 3.2 Take a matrix element of the completeness relation in Eq. (3.97) between the states $|\xi'\rangle$ and $|\xi''\rangle$, and show that one obtains the correct expression for $\langle \xi' | \xi'' \rangle$.

Solution to Problem 3.2

The completeness relation for the eigenstates of position in Eq. (3.97) states that

$$\int d\xi \, |\xi\rangle\langle\xi| = 1_{\rm op}$$

The continuum orthonormality statement for these states is given in Eqs. (3.83) as

$$\langle \xi' | \xi \rangle = \delta(\xi' - \xi)$$

The matrix element of the completeness relation between the states $|\xi'\rangle$ and $|\xi''\rangle$ then gives

$$\langle \xi'' | \xi' \rangle = \int d\xi \, \langle \xi'' | \xi \rangle \langle \xi | \xi' \rangle$$
$$= \int d\xi \, \delta(\xi'' - \xi)\delta(\xi - \xi') = \delta(\xi'' - \xi')$$

which is the proper result.

Problem 3.3 Show that the Schrödinger equation in the momentum representation in Eq. (3.121) is obtained as the Fourier transform of the Schrödinger equation in the coordinate representation in Eq. (3.113).

Solution to Problem 3.3

The Schrödinger equation in the coordinate representation is

$$\left[-\frac{\hbar^2}{2m}\nabla^2 + V(\mathbf{x}) \right] \psi(\mathbf{x}) = E\psi(\mathbf{x})$$

Introduce the Fourier transform of the potential

$$V(\mathbf{x}) = \int \frac{d^3q}{(2\pi)^3} e^{i\mathbf{q}\cdot\mathbf{x}} \tilde{V}(\mathbf{q})$$

Also introduce the Fourier transform of the wave function

$$A(\mathbf{k}) = \int d^3x\, e^{-i\mathbf{k}\cdot\mathbf{x}}\, \psi(\mathbf{x})$$

Now perform the operation $\int d^3x\, e^{-i\mathbf{k}\cdot\mathbf{x}}$ on the Schrödinger equation in the coordinate representation to obtain

$$-\frac{\hbar^2}{2m}\int d^3x\, e^{-i\mathbf{k}\cdot\mathbf{x}}\nabla^2\psi(\mathbf{x}) + \int d^3x\, e^{-i\mathbf{k}\cdot\mathbf{x}}\int \frac{d^3q}{(2\pi)^3} e^{i\mathbf{q}\cdot\mathbf{x}}\tilde{V}(\mathbf{q})\,\psi(\mathbf{x}) = EA(\mathbf{k})$$

The hermiticity of ∇^2 allows the first term to be written

$$-\frac{\hbar^2}{2m}\int d^3x\, e^{-i\mathbf{k}\cdot\mathbf{x}}\nabla^2\psi(\mathbf{x}) = \frac{\hbar^2 \mathbf{k}^2}{2m}A(\mathbf{k})$$

The second term takes the form

$$\int d^3x\, e^{-i\mathbf{k}\cdot\mathbf{x}}\int \frac{d^3q}{(2\pi)^3} e^{i\mathbf{q}\cdot\mathbf{x}}\tilde{V}(\mathbf{q})\,\psi(\mathbf{x}) = \int \frac{d^3q}{(2\pi)^3}\tilde{V}(\mathbf{q})A(\mathbf{k}-\mathbf{q})$$

$$= \int \frac{d^3k'}{(2\pi)^3}\tilde{V}(\mathbf{k}-\mathbf{k}')A(\mathbf{k}')$$

where the second line arises from a change of variable to $\mathbf{k}' \equiv \mathbf{k} - \mathbf{q}$.[1]

---

[1]Note that $\int d^3q = \int d^3k'$; note also that the final equation is independent of the normalization of $A(\mathbf{k})$.

The Fourier transform of the Schrödinger equation in the coordinate representation therefore becomes

$$\left(E - \frac{\hbar^2 \mathbf{k}^2}{2m}\right) A(\mathbf{k}) = \int \frac{d^3 k'}{(2\pi)^3} \tilde{V}(\mathbf{k} - \mathbf{k}') A(\mathbf{k}')$$

This is the large-volume limit of the Schrödinger equation in the momentum representation in Eq. (3.121).

Problem 3.4 Use completeness to derive the following relations in the coordinate and momentum representations

$$\langle \boldsymbol{\xi} | \hat{\mathbf{p}} | \boldsymbol{\xi}' \rangle = \frac{\hbar}{i} \boldsymbol{\nabla}_\xi \langle \boldsymbol{\xi} | \boldsymbol{\xi}' \rangle \qquad ; \qquad \langle \mathbf{k} | \hat{\mathbf{x}} | \mathbf{k}' \rangle = i\hbar \boldsymbol{\nabla}_p \langle \mathbf{k} | \mathbf{k}' \rangle$$

Solution to Problem 3.4

Use the completeness relation for the eigenstates of momentum

$$\sum_{\mathbf{k}} |\mathbf{k}\rangle\langle\mathbf{k}| = \hat{1}$$

and the inner product in Eq. (3.109) to write

$$\langle \boldsymbol{\xi} | \hat{\mathbf{p}} | \boldsymbol{\xi}' \rangle = \sum_{\mathbf{k}} \langle \boldsymbol{\xi} | \hat{\mathbf{p}} | \mathbf{k} \rangle \langle \mathbf{k} | \boldsymbol{\xi}' \rangle$$

$$= \frac{1}{L^3} \sum_{\mathbf{k}} \hbar \mathbf{k}\, e^{i(\mathbf{k}-\mathbf{k}')\cdot\boldsymbol{\xi}} = \frac{\hbar}{i} \boldsymbol{\nabla}_\xi \left[\frac{1}{L^3} \sum_{\mathbf{k}} e^{i(\mathbf{k}-\mathbf{k}')\cdot\boldsymbol{\xi}}\right]$$

$$= \frac{\hbar}{i} \boldsymbol{\nabla}_\xi \langle \boldsymbol{\xi} | \boldsymbol{\xi}' \rangle$$

In a similar fashion, with the completeness relation for the eigenstates of position

$$\int d^3\xi\, |\boldsymbol{\xi}\rangle\langle\boldsymbol{\xi}| = \hat{1}$$

one obtains

$$\langle \mathbf{k} | \hat{\mathbf{x}} | \mathbf{k}' \rangle = \int d^3\xi\, \langle \mathbf{k} | \hat{\mathbf{x}} | \boldsymbol{\xi} \rangle \langle \boldsymbol{\xi} | \mathbf{k}' \rangle$$

$$= \frac{1}{L^3} \int d^3\xi\, \boldsymbol{\xi}\, e^{-i(\mathbf{k}-\mathbf{k}')\cdot\boldsymbol{\xi}} = i\boldsymbol{\nabla}_k \left[\frac{1}{L^3} \int d^3\xi\, e^{-i(\mathbf{k}-\mathbf{k}')\cdot\boldsymbol{\xi}}\right]$$

$$= i\hbar \boldsymbol{\nabla}_p \langle \mathbf{k} | \mathbf{k}' \rangle$$

where $\hbar \mathbf{k} = \mathbf{p}$.

Problem 3.5 (a) Consider the harmonic oscillator, and verify the example of Ehrenfest's theorem in Eqs. (3.163);[2]

$$\frac{d}{dt}\langle\Psi(t)|\hat{x}|\Psi(t)\rangle = \langle\Psi(t)|\frac{i}{\hbar}[\hat{H},\hat{x}]|\Psi(t)\rangle = \frac{1}{m}\langle\Psi(t)|\hat{p}|\Psi(t)\rangle$$

(b) Then show, in addition, that

$$\frac{d}{dt}\langle\Psi(t)|\hat{p}|\Psi(t)\rangle = \langle\Psi(t)|\frac{i}{\hbar}[\hat{H},\hat{p}]|\Psi(t)\rangle = -k\langle\Psi(t)|\hat{x}|\Psi(t)\rangle$$

(c) Use these results to verify Newton's second law for the classical limit of the motion of a particle in an oscillator potential.

Solution to Problem 3.5

(a) The operator $\hat{x}$ has no explicit time dependence. Thus, following the analysis in Eqs. (3.169)–(3.170), one has

$$\frac{d}{dt}\langle\Psi(t)|\hat{x}|\Psi(t)\rangle = \langle\frac{\partial}{\partial t}\Psi(t)|\hat{x}|\Psi(t)\rangle + \langle\Psi(t)|\hat{x}\frac{\partial}{\partial t}\Psi(t)\rangle$$

$$= -\frac{1}{i\hbar}\langle\hat{H}\Psi(t)|\hat{x}|\Psi(t)\rangle + \frac{1}{i\hbar}\langle\Psi(t)|\hat{x}\hat{H}|\Psi(t)\rangle$$

The hamiltonian is hermitian, and so

$$\frac{d}{dt}\langle\Psi(t)|\hat{x}|\Psi(t)\rangle = \frac{i}{\hbar}\langle\Psi(t)|[\hat{H},\hat{x}]|\Psi(t)\rangle$$

The hamiltonian for the harmonic oscillator is

$$\hat{H} = \frac{1}{2m}\hat{p}^2 + \frac{1}{2}k\hat{x}^2$$

Therefore

$$[\hat{H},\hat{x}] = \frac{1}{2m}[\hat{p}^2,\hat{x}]$$

$$= \frac{1}{2m}\{\hat{p}[\hat{p},\hat{x}] + [\hat{p},\hat{x}]\hat{p}\} = \frac{\hbar}{i}\frac{\hat{p}}{m}$$

Hence

$$\frac{d}{dt}\langle\Psi(t)|\hat{x}|\Psi(t)\rangle = \langle\Psi(t)|\frac{i}{\hbar}[\hat{H},\hat{x}]|\Psi(t)\rangle = \frac{1}{m}\langle\Psi(t)|\hat{p}|\Psi(t)\rangle$$

(b) In a similar fashion

$$\frac{d}{dt}\langle\Psi(t)|\hat{p}|\Psi(t)\rangle = \frac{i}{\hbar}\langle\Psi(t)|[\hat{H},\hat{p}]|\Psi(t)\rangle$$

---

[2]Compare the discussion in Eqs. (3.169)–(3.171).

In this case, the commutator is

$$[\hat{H}, \hat{p}] = \frac{k}{2}[\hat{x}^2, \hat{p}]$$

$$= \frac{k}{2}\left\{\hat{x}[\hat{x}, \hat{p}] + [\hat{x}, \hat{p}]\hat{x}\right\} = -\frac{\hbar}{i}k\hat{x}$$

Therefore

$$\frac{d}{dt}\langle\Psi(t)|\hat{p}|\Psi(t)\rangle = \langle\Psi(t)|\frac{i}{\hbar}[\hat{H}, \hat{p}]|\Psi(t)\rangle = -k\langle\Psi(t)|\hat{x}|\Psi(t)\rangle$$

(c) In the classical limit, one has a wave packet that is very well localized in both position and momentum, so that

$$\langle\Psi(t)|\hat{x}|\Psi(t)\rangle \approx x(t)$$
$$\langle\Psi(t)|\hat{p}|\Psi(t)\rangle \approx p(t)$$

The above results then become

$$\frac{dx(t)}{dt} = \frac{p(t)}{m}$$
$$\frac{dp(t)}{dt} = -kx(t)$$

These are just Hamilton's equations (Newton's law) for the simple harmonic oscillator.

Problem 3.6 (a) Consider the following operator $\hat{F}(\lambda) \equiv e^{i\lambda\hat{A}}\hat{B}e^{-i\lambda\hat{A}}$. Show

$$\frac{d^n\hat{F}(\lambda)}{d\lambda^n}\bigg|_{\lambda=0} = i^n[\hat{A}, \cdots, [\hat{A}, [\hat{A}, \hat{B}]]\cdots] \qquad ; n \text{ terms}$$

(b) Make a Taylor series expansion of $\hat{F}(\lambda)$ in λ, and then set $\lambda = 1$ to verify Eq. (3.122).

Solution to Problem 3.6

(a) Take one derivative of the operator $\hat{F}(\lambda) \equiv e^{i\lambda\hat{A}}\hat{B}e^{-i\lambda\hat{A}}$ with respect to λ

$$\frac{d}{d\lambda}\hat{F}(\lambda) = \frac{d}{d\lambda}e^{i\lambda\hat{A}}\hat{B}e^{-i\lambda\hat{A}} = e^{i\lambda\hat{A}}i[\hat{A}, \hat{B}]e^{-i\lambda\hat{A}}$$

Take a second derivative

$$\frac{d^2}{d\lambda^2}\hat{F}(\lambda) = e^{i\lambda\hat{A}}i^2[\hat{A},[\hat{A},\hat{B}]]e^{-i\lambda\hat{A}}$$

Repeat this process, and then set $\lambda = 0$ to obtain

$$\left.\frac{d^n\hat{F}(\lambda)}{d\lambda^n}\right|_{\lambda=0} = i^n[\hat{A},\cdots,[\hat{A},[\hat{A},\hat{B}]]\cdots] \qquad ; n \text{ terms}$$

(b) A Taylor series expansion of $\hat{F}(\lambda)$ about the origin gives[3]

$$\hat{F}(\lambda) = \sum_n \frac{\lambda^n}{n!}\left.\frac{d^n\hat{F}(\lambda)}{d\lambda^n}\right|_{\lambda=0}$$

$$= \sum_n \frac{(i\lambda)^n}{n!}[\hat{A},\cdots,[\hat{A},[\hat{A},\hat{B}]]\cdots]$$

Now set $\lambda = 1$, with the result

$$\hat{F}(1) = e^{i\hat{A}}\hat{B}e^{-i\hat{A}} = \sum_n \frac{i^n}{n!}[\hat{A},\cdots,[\hat{A},[\hat{A},\hat{B}]]\cdots]$$

This is Eq. (3.122).

Problem 3.7 Consider the harmonic oscillator in the Heisenberg picture. Use Eqs. (3.148), (3.158), and (3.122) to show that

$$\hat{a}_H(t) = e^{i\hat{H}t/\hbar}\,\hat{a}\,e^{-i\hat{H}t/\hbar} = \hat{a}\,e^{-i\omega_0 t}$$
$$\hat{a}_H^\dagger(t) = e^{i\hat{H}t/\hbar}\,\hat{a}^\dagger e^{-i\hat{H}t/\hbar} = \hat{a}^\dagger e^{i\omega_0 t}$$

Solution to Problem 3.7

Equation (3.122) states that

$$e^{i\hat{A}}\hat{B}e^{-i\hat{A}} \equiv \hat{B} + i[\hat{A},\hat{B}] + \frac{i^2}{2!}[\hat{A},[\hat{A},\hat{B}]] + \frac{i^3}{3!}[\hat{A},[\hat{A},[\hat{A},\hat{B}]]] + \cdots$$

The hamiltonian for the simple harmonic oscillator is given by Eq. (3.158)

$$\hat{H} = \hbar\omega_0\left(\hat{N} + \frac{1}{2}\right)$$

---

[3]If you are more comfortable with functions, take matrix elements of the operator.

where $\hat{N}$ is the number operator, satisfying the commutation relations in Eqs. (3.148)

$$[\hat{N}, \hat{a}] = -\hat{a}$$
$$[\hat{N}, \hat{a}^\dagger] = \hat{a}^\dagger$$

Substitution of these relations in the above allows us to calculate, for example, the destruction operator in the Heisenberg picture

$$e^{i\hat{H}t/\hbar}\hat{a}e^{-i\hat{H}t/\hbar} = \hat{a} + \frac{it}{\hbar}[\hat{H}, \hat{a}] + \left(\frac{it}{\hbar}\right)^2 \frac{1}{2!}[\hat{H}, [\hat{H}, \hat{a}]]$$
$$+ \left(\frac{it}{\hbar}\right)^3 \frac{1}{3!}[\hat{H}, [\hat{H}, [\hat{H}, \hat{a}]]] + \cdots$$

Evaluation of the required commutators gives

$$e^{i\hat{H}t/\hbar}\hat{a}e^{-i\hat{H}t/\hbar} = \hat{a} + (-it\omega_0)\hat{a} + \frac{(-it\omega_0)^2}{2!}\hat{a} + \frac{(-it\omega_0)^3}{3!}\hat{a} + \cdots$$
$$= \hat{a}e^{-i\omega_0 t}$$

An analogous calculation gives

$$e^{i\hat{H}t/\hbar}\hat{a}^\dagger e^{-i\hat{H}t/\hbar} = \hat{a}^\dagger e^{i\omega_0 t}$$

Problem 3.8 (a) Start from Eqs. (3.161), use the results in Prob. 3.7, and show by explicit differentiation with respect to time that for the harmonic oscillator in the Heisenberg picture

$$m\frac{d\hat{x}_H(t)}{dt} = \hat{p}_H(t) \qquad ; \qquad \frac{d\hat{p}_H(t)}{dt} = -k\,\hat{x}_H(t)$$

Compare with the results in Prob. 3.5;

(b) Show that in the Heisenberg picture, it is the *equal-time* commutation relation that yields the canonical value

$$[\hat{p}_H(t), \hat{x}_H(t')]_{t=t'} = \frac{\hbar}{i}$$

Solution to Problem 3.8

(a) We start from Eqs. (3.161) and use the results of Prob. 3.7 to express the position and momentum operators for the oscillator in the Heisenberg picture as

$$\hat{x}_H(t) = e^{i\hat{H}t/\hbar}\hat{x}e^{-i\hat{H}t/\hbar} = \left(\frac{\hbar}{2m\omega_0}\right)^{1/2}\left(\hat{a}e^{-i\omega_0 t} + \hat{a}^\dagger e^{i\omega_0 t}\right)$$

$$\hat{p}_H(t) = e^{i\hat{H}t/\hbar}\hat{p}e^{-i\hat{H}t/\hbar} = -im\omega_0\left(\frac{\hbar}{2m\omega_0}\right)^{1/2}\left(\hat{a}e^{-i\omega_0 t} - \hat{a}^\dagger e^{i\omega_0 t}\right)$$

The time derivative may now be explicitly evaluated to arrive at

$$m\frac{d\hat{x}_H(t)}{dt} = \hat{p}_H(t) \qquad ; \qquad \frac{d\hat{p}_H(t)}{dt} = -k\,\hat{x}_H(t)$$

where $k \equiv m\omega_0^2$. These are exactly the same relations obtained for the expectation values in the Schrödinger picture in Prob. 3.5.

(b) It is the $(\hat{a}, \hat{a}^\dagger)$ that are the operators in the above expressions, and they satisfy Eqs. (3.139). The equal-time commutator of the momentum and position in the Heisenberg picture then reproduces the canonical value

$$[\hat{p}_H(t), \hat{x}_H(t')]_{t=t'} = -\frac{i\hbar}{2}\left([\hat{a}, \hat{a}^\dagger] - [\hat{a}^\dagger, \hat{a}]\right) = \frac{\hbar}{i}$$

for all time t.

Problem 3.9 For clarity, the discussion of measurements has essentially been confined to one dimension. The simplest extension to three dimensions is to eigenstates of momentum with periodic boundary conditions. Here the eigenstates are the direct product $|\mathbf{k}\rangle = |k_x\rangle|k_y\rangle|k_z\rangle$, with $k_i = 2\pi n_i/L$ and $n_i = 0, \pm 1, \pm 2, \cdots$, and the statement of completeness factors as

$$\sum_{\mathbf{k}}|\mathbf{k}\rangle\langle\mathbf{k}| = \sum_{k_x}\sum_{k_y}\sum_{k_z}|k_x\rangle|k_y\rangle|k_z\rangle\langle k_x|\langle k_y|\langle k_z| = \hat{1}$$

With the central-force problem in three dimensions, the situation is a little more complicated. The maximal set of mutually commuting operators is taken as $(\hat{H}, \hat{\mathbf{L}}^2, \hat{L}_z)$, and the eigenstates are $|nlm\rangle = |nl\rangle|lm\rangle$ where[4]

$$\hat{\mathbf{L}}^2|lm\rangle = l(l+1)|lm\rangle \qquad ; \qquad \hat{L}_z|lm\rangle = m|lm\rangle \qquad ; \qquad \hat{H}|nl\rangle = E_{nl}|nl\rangle$$

These states satisfy the orthonormality relations $\langle lm|l'm'\rangle = \delta_{ll'}\delta_{mm'}$ and $\langle nl|n'l\rangle = \delta_{nn'}$. The completeness statement is

$$\sum_{nlm}|nlm\rangle\langle nlm| = \sum_{nlm}|nl\rangle|lm\rangle\langle nl|\langle lm| = \hat{1}$$

---

[4]The corresponding wave functions are $\langle\mathbf{x}|nlm\rangle = \langle r|nl\rangle\langle\theta\phi|lm\rangle = R_{nl}(r)Y_{lm}(\theta, \phi)$.

(a) Take the matrix element $\langle \mathbf{x}| \cdots |\mathbf{x}'\rangle$ of this last relation and reproduce the completeness statement in the coordinate representation

$$\sum_{nlm} R_{nl}(r)R_{nl}^{\star}(r')Y_{lm}(\theta,\phi)Y_{lm}^{\star}(\theta',\phi') = \frac{1}{r^2\sin\theta}\delta(r-r')\delta(\theta-\theta')\delta(\phi-\phi')$$

State carefully how the sums are to be carried out;

(b) Let $\hat{L}_{\pm} \equiv \hat{L}_x \pm i\hat{L}_y$. Show

$$\langle nlm|[\hat{\mathbf{L}}^2,\hat{L}_{\pm}]|n'l'm'\rangle = [l(l+1)-l'(l'+1)]\delta_{nn'}\langle lm|\hat{L}_{\pm}|l'm'\rangle = 0$$

Hence conclude that the matrices $\underline{L}_{\pm}$ are block-diagonal in this basis.

Solution to Problem 3.9

(a) Take matrix elements of the completeness relation between eigenstates of position in the central-force problem

$$\sum_{nlm}\langle \mathbf{x}|nlm\rangle\langle nlm|\mathbf{x}'\rangle = \langle \mathbf{x}|\mathbf{x}'\rangle$$

On the r.h.s. use the orthonormality of the states

$$\langle \mathbf{x}|\mathbf{x}'\rangle = \delta^{(3)}(\mathbf{x}-\mathbf{x}')$$
$$= \frac{1}{r^2\sin\theta}\delta(r-r')\delta(\theta-\theta')\delta(\phi-\phi')$$

Here the three-dimensional delta-function has been written in spherical coordinates where the volume element is $d^3x = r^2\sin\theta\,drd\theta d\phi$.

On the l.h.s. of the original expression, use the decomposition of the state vector indicated in the problem to write

$$\sum_{nlm}\langle \mathbf{x}|nlm\rangle\langle nlm|\mathbf{x}'\rangle = \sum_{nlm}\langle r|nl\rangle\langle \theta\phi|lm\rangle\langle nl|r'\rangle\langle lm|\theta'\phi'\rangle$$
$$= \sum_{nlm} R_{nl}(r)R_{nl}^{\star}(r')Y_{lm}(\theta,\phi)Y_{lm}^{\star}(\theta',\phi')$$

Therefore

$$\sum_{nlm} R_{nl}(r)R_{nl}^{\star}(r')Y_{lm}(\theta,\phi)Y_{lm}^{\star}(\theta',\phi') = \frac{1}{r^2\sin\theta}\delta(r-r')\delta(\theta-\theta')\delta(\phi-\phi')$$

With respect to the order of the sums:

- The radial wave functions are the solutions to the Sturm-Liouville problem in Eq. (2.199) for each l, and therefore they are complete *for each l.*[5] Hence

$$\sum_n \langle r|nl\rangle\langle nl|r'\rangle = \sum_n R_{nl}(r)R_{nl}^\star(r')$$

$$= \frac{1}{r^2}\delta(r-r')$$

- The associated Legendre polynomials are solutions to the Sturm-Liouville problem in Eq. (2.153) *for each m.*[6] Thus

$$\sum_l \langle\theta\phi|lm\rangle\langle lm|\theta'\phi'\rangle = \sum_l Y_{lm}(\theta,\phi)Y_{lm}^\star(\theta',\phi')$$

$$= \frac{1}{2\pi}e^{im(\phi-\phi')}\frac{1}{\sin\theta}\delta(\theta-\theta')$$

- The azimuthal wave functions form a complex Fourier series, and are complete in ϕ. Therefore

$$\sum_{ml} \langle\theta\phi|lm\rangle\langle lm|\theta'\phi'\rangle = \sum_{ml} Y_{lm}(\theta,\phi)Y_{lm}^\star(\theta',\phi')$$

$$= \frac{1}{\sin\theta}\delta(\theta-\theta')\delta(\phi-\phi')$$

- Now the sum over (l,m) can be carried out in either order, provided proper care is taken with respect to the domain of summation.

(b) The operators $\hat{L}_\pm \equiv \hat{L}_x \pm i\hat{L}_y$ and $\hat{\mathbf{L}}^2$ act only on the states $|lm\rangle$, and these states are eigenstates of $\hat{\mathbf{L}}^2$ with eigenvalues $l(l+1)$

$$\hat{\mathbf{L}}^2|nlm\rangle = \hat{\mathbf{L}}^2|nl\rangle|lm\rangle = l(l+1)|nlm\rangle$$

Furthermore, these operators commute

$$[\hat{\mathbf{L}}^2, \hat{L}_\pm] = 0$$

so that any matrix element vanishes. Therefore

$$\langle nlm|[\hat{\mathbf{L}}^2, \hat{L}_\pm]|n'l'm'\rangle = [l(l+1)-l'(l'+1)]\langle nlm|\hat{L}_\pm|n'l'm'\rangle$$

$$= 0$$

---

[5]This assumes that $V(r)$ is well-behaved.
[6]Here $\alpha = l$.

It follows that the matrix element on the r.h.s. is only non-zero if $l = l'$. Thus, in the notation of the problem

$$\langle nlm|\hat{L}_{\pm}|n'l'm'\rangle = \delta_{ll'}\langle nl|n'l\rangle\langle lm|\hat{L}_{\pm}|lm'\rangle$$

Now use the orthonormality of the radial eigenfunctions for each l

$$\langle nl|n'l\rangle = \delta_{nn'}$$

We conclude that the matrices $\underline{L}_{\pm}$ are block-diagonal in this basis

$$\langle nlm|\hat{L}_{\pm}|n'l'm'\rangle = \delta_{nn'}\delta_{ll'}\langle lm|\hat{L}_{\pm}|lm'\rangle$$

Problem 3.10 Show that both the trace and determinant of the hamiltonian matrix H_{mn} are invariant under a unitary transformation.

Solution to Problem 3.10

Let $\underline{U}$ be a unitary matrix satisfying

$$\underline{U}^{\dagger} = \underline{U}^{-1}$$

Under a unitary transformation, the hamiltonian matrix gets transformed into

$$\underline{H}' = \underline{U}^{\dagger}\underline{H}\,\underline{U} = \underline{U}^{-1}\underline{H}\,\underline{U}$$

The determinant of a product of matrices is the product of determinants

$$\det \underline{A}\,\underline{B}\,\underline{C} = \det \underline{A}\,\det \underline{B}\,\det \underline{C}$$

Hence

$$\begin{aligned}
\det \underline{H}' &= \det \underline{U}^{-1}\det \underline{H}\,\det \underline{U} \\
&= \det \left(\underline{U}^{-1}\underline{U}\right)\det \underline{H} \\
&= \det \underline{H}
\end{aligned}$$

where we have used $\det \underline{1} = 1$.

The trace of a product of matrices is invariant under a cyclic permutation

$$\operatorname{tr} \underline{A}\,\underline{B}\,\underline{C} = \operatorname{tr} \underline{B}\,\underline{C}\,\underline{A} = \operatorname{tr} \underline{C}\,\underline{A}\,\underline{B}$$

Therefore

$$\operatorname{tr} \underline{H}' = \operatorname{tr} \underline{U}\,\underline{U}^{-1}\underline{H}$$
$$= \operatorname{tr} \underline{H}$$

PART 2
Applications of Quantum Mechanics

Chapter 4

Approximation Methods for Bound States

Problem 4.1 (a) Write the energy functional in the first of Eqs. (4.32) in spherical coordinates with a $\psi(r)$ as

$$\mathcal{E} = 4\pi \int_0^\infty r^2 dr \, \psi^\star \left[-\frac{\hbar^2}{2m} \frac{1}{r^2} \frac{d}{dr} r^2 \frac{d}{dr} + V(r) \right] \psi$$

Now partially integrate the first term with the wave function in Eq. (4.30) to verify the second of Eqs. (4.32)

$$\mathcal{E} = 4\pi \int_0^\infty r^2 dr \left[\frac{\hbar^2}{2m} \left| \frac{d\psi}{dr} \right|^2 + V(r)|\psi|^2 \right]$$

(b) Verify the numerical results in Fig. 4.3 and Eqs. (4.36).

Solution to Problem 4.1

(a) We are to evaluate the energy functional

$$\mathcal{E} = \int \psi^\star \left[-\frac{\hbar^2}{2m} \nabla^2 + V(r) \right] \psi \, d^3 r$$

with the normalized wave function

$$\psi(r) = \left[\frac{(2\alpha)^3}{8\pi} \right]^{1/2} e^{-\alpha r}$$

The radial part of the laplacian is

$$\nabla^2 \doteq \frac{1}{r^2} \frac{d}{dr} r^2 \frac{d}{dr}$$

83

Substitute this into the energy functional

$$\mathcal{E} = 4\pi \int_0^\infty r^2 dr \, \psi^\star \left[-\frac{\hbar^2}{2m} \frac{1}{r^2} \frac{d}{dr} r^2 \frac{d}{dr} + V(r) \right] \psi$$

$$= 4\pi \int_0^\infty dr \, \psi^\star \left[-\frac{\hbar^2}{2m} \frac{d}{dr} r^2 \frac{d}{dr} + r^2 V(r) \right] \psi$$

Now carry out a partial integration on the kinetic-energy term

$$\int_0^\infty dr \, \psi^\star \left[-\frac{d}{dr} r^2 \frac{d}{dr} \right] \psi = - \left[\psi^\star r^2 \frac{d}{dr} \psi \right]_0^\infty + \int_0^\infty r^2 dr \left| \frac{d\psi}{dr} \right|^2$$

The first term explicitly vanishes at both limits, and therefore

$$\mathcal{E} = 4\pi \int_0^\infty r^2 dr \left[\frac{\hbar^2}{2m} \left| \frac{d\psi}{dr} \right|^2 + V(r)|\psi|^2 \right]$$

(b) Done in text.

Problem 4.2 Assume the potential is bounded from below. Show the energy functional satisfies $\mathcal{E} \geq V_{\min}$ where $V_{\min}$ is the minimum value of V.

Solution to Problem 4.2

The energy functional is given by

$$\mathcal{E} = \frac{\int d^3x \, \psi^\star \left[-\hbar^2 \nabla^2 / 2m + V(x) \right] \psi}{\int d^3x \, |\psi|^2}$$

As in Eq. (4.32), the hermiticity of $\mathbf{p} = (\hbar/i)\nabla$ allows this to be written as

$$\mathcal{E} = \frac{\int d^3x \left[(\hbar^2/2m)|\nabla\psi|^2 + V(x)|\psi|^2 \right]}{\int d^3x \, |\psi|^2}$$

The kinetic energy term is evidently positive definite, and hence

$$\mathcal{E} \geq \frac{\int d^3x \, V(x)|\psi|^2}{\int d^3x \, |\psi|^2}$$

Now write this as

$$\mathcal{E} \geq \frac{\int d^3x \left[V_{\min} + V(x) - V_{\min} \right] |\psi|^2}{\int d^3x \, |\psi|^2}$$

$$= V_{\min} + \frac{\int d^3x \left[V(x) - V_{\min} \right] |\psi|^2}{\int d^3x \, |\psi|^2}$$

Since the last term in the second line is again positive definite, we have the desired inequality

$$\mathcal{E} \geq V_{\min}$$

Problem 4.3 Show through order H' that the wave functions $\psi_n^{(1)}$ in non-degenerate perturbation theory with an hermitian H' form an orthonormal system

$$\langle \psi_n^{(1)} | \psi_m^{(1)} \rangle = \delta_{nm} + O(H'^2) \qquad ; \text{ orthonormal}$$

Solution to Problem 4.3

The wave function in first-order non-degenerate perturbation theory is given in Eq. (4.58) as

$$\psi_n^{(1)}(x) = \phi_n(x) + \sum_{p \neq n} \phi_p(x) \frac{\langle \phi_p | H' | \phi_n \rangle}{E_n^0 - E_p^0}$$

Now compute the inner product $\langle \psi_n^{(1)} | \psi_m^{(1)} \rangle$

$$\langle \psi_n^{(1)} | \psi_m^{(1)} \rangle = \langle \phi_n | \phi_m \rangle + \sum_{p \neq m} \frac{\langle \phi_n | \phi_p \rangle \langle \phi_p | H' | \phi_m \rangle}{E_m^0 - E_p^0}$$
$$+ \sum_{p \neq n} \frac{\langle \phi_p | \phi_m \rangle \langle \phi_p | H' | \phi_n \rangle^\star}{E_n^0 - E_p^0} + O\left(H'^2\right).$$

The orthonormality of the unperturbed wave functions $\langle \phi_n | \phi_m \rangle = \delta_{mn}$, and the hermiticity of H', then yield

$$\langle \psi_n^{(1)} | \psi_m^{(1)} \rangle = \delta_{mn} + \frac{\langle \phi_n | H' | \phi_m \rangle}{E_m^0 - E_n^0} + \frac{\langle \phi_n | H' | \phi_m \rangle}{E_n^0 - E_m^0} + O\left(H'^2\right)$$

The second and third terms cancel identically,[1] and this is the desired result

$$\langle \psi_n^{(1)} | \psi_m^{(1)} \rangle = \delta_{mn} + O\left(H'^2\right)$$

Problem 4.4 (a) Show the second-order wave function in non-

---

[1] The restrictions on the sums ensure that these terms indeed disappear when $n = m$.

degenerate perturbation theory is given by

$$\psi_n^{(2)} = \phi_n + \sum_{m \neq n} \phi_m \frac{\langle \phi_m | H' | \phi_n \rangle}{E_n^0 - E_m^0} \left[1 - \frac{\langle \phi_n | H' | \phi_n \rangle}{E_n^0 - E_m^0} \right]$$
$$+ \sum_{m \neq n} \sum_{p \neq n} \phi_m \frac{\langle \phi_m | H' | \phi_p \rangle \langle \phi_p | H' | \phi_n \rangle}{(E_n^0 - E_m^0)(E_n^0 - E_p^0)}$$

(b) Verify that this expression is well-defined if the system is non-degenerate [see Eq. (4.59)].

Solution to Problem 4.4

(a) The second-order expression for the wave function $\psi_n^{(2)}$ in non-degenerate perturbation theory can be obtained from Eqs. (4.56) in two steps:

(1) Use the first-order expression for the wave function in Eqs. (4.58)[2] in the matrix element on the r.h.s. of the equation for ψ;
(2) Then substitute the first-order expression for the energy in Eqs. (4.58) in the denominator of the equation for ψ, and expand in H'.

If the terms up through $O(H'^2)$ are retained, the result is

$$\psi_n^{(2)} = \phi_n + \sum_{m \neq n} \phi_m \frac{\langle \phi_m | H' | \phi_n \rangle}{E_n^0 - E_m^0} \left[1 - \frac{\langle \phi_n | H' | \phi_n \rangle}{E_n^0 - E_m^0} \right]$$
$$+ \sum_{m \neq n} \sum_{p \neq n} \phi_m \frac{\langle \phi_m | H' | \phi_p \rangle \langle \phi_p | H' | \phi_n \rangle}{(E_n^0 - E_m^0)(E_n^0 - E_p^0)}$$

(b) It follows from Eq. (4.59) that with a non-degenerate spectrum, if $n \neq l$ then $E_n^{(0)} \neq E_l^{(0)}$. Hence all the energy denominators in the above expression are non-zero, and $\psi_n^{(2)}$ is well-defined and truly of $O(H'^2)$.

Problem 4.5 Prove that the second-order energy shift always lowers the energy of the ground state in non-degenerate perturbation theory.

Solution to Problem 4.5

The second-order contribution to the energy shift in non-degenerate

---

[2]With a dummy summation index $m \to p$.

perturbation theory is given by Eq. (4.60) as

$$\delta E_n^{(2)} = \sum_{m \neq n} \frac{|\langle \phi_m | H' | \phi_n \rangle|^2}{E_n^0 - E_m^0}$$

If ϕ_n is the ground state, then all the contributions to the sum are negative definite since $E_m^0 - E_n^0 > 0$, and hence the second-order energy shift always lowers the energy of the ground state in non-degenerate perturbation theory.

Problem 4.6 Reproduce the λ^2 term in Eq. (4.77) using second-order non-degenerate perturbation theory.

Solution to Problem 4.6

The second-order energy shift in non-degenerate perturbation theory is given by the above

$$\delta E_n^{(2)} = \sum_{m \neq n} \frac{|\langle \phi_m | H' | \phi_n \rangle|^2}{E_n^0 - E_m^0}$$

Here, the interaction hamiltonian is

$$H' \equiv \lambda \left(\frac{1}{2} m_0 \omega_0^2 \right) x^2$$

with the mass denoted by m_0. It is most convenient to express the coordinate operator in terms of creation and destruction operators [see Eqs. (3.161)]

$$\hat{x} = \left(\frac{\hbar}{2 m_0 \omega_0} \right)^{1/2} (\hat{a} + \hat{a}^\dagger)$$

Then

$$\delta E_n^{(2)} = \left[\lambda \left(\frac{1}{2} m_0 \omega_0^2 \right) \right]^2 \left(\frac{\hbar}{2 m_0 \omega_0} \right)^2 \sum_{m \neq n} \frac{|\langle m | (\hat{a} + \hat{a}^\dagger)(\hat{a} + \hat{a}^\dagger) | n \rangle|^2}{E_n^0 - E_m^0}$$

There are two possibilities in the sum, $|m\rangle = |n + 2\rangle$ and $|m\rangle = |n - 2\rangle$. Therefore

$$\delta E_n^{(2)} = \frac{\lambda^2}{16} (\hbar \omega_0)^2 \left[\frac{|\langle n + 2 | \hat{a}^\dagger \hat{a}^\dagger | n \rangle|^2}{-2 \hbar \omega_0} + \frac{|\langle n - 2 | \hat{a} \hat{a} | n \rangle|^2}{2 \hbar \omega_0} \right]$$

Now use the properties of the creation and destruction operators

$$\hat{a}^\dagger | n \rangle = \sqrt{n + 1} \, | n + 1 \rangle \qquad ; \, \hat{a} | n \rangle = \sqrt{n} \, | n - 1 \rangle$$

to obtain

$$\delta E_n^{(2)} = \frac{\lambda^2}{32} \hbar \omega_0 \left[-(n+1)(n+2) + n(n-1)\right]$$

$$= -\frac{\lambda^2}{8} \hbar \omega_0 \left(n + \frac{1}{2}\right)$$

$$= -\frac{\lambda^2}{8} E_n^0$$

This result now reproduces the λ^2 term in Eq. (4.77).

Problem 4.7 Show that the Coulomb repulsion in Eqs. (4.92) and (4.95) has the following interpretation. It is the Coulomb potential provided by the second electron at a radial distance ρ_1 integrated over the charge density of the first electron in a shell of thickness $d\rho_1$.

Solution to Problem 4.7

The Coulomb interaction energy of two electrons in a $1s$-orbit is

$$V_C = \int d^3 r_1 \int d^3 r_2 \, \rho(r_1) \frac{e^2}{|\mathbf{r}_1 - \mathbf{r}_2|} \rho(r_2)$$

$$\rho(r) = |\psi_{1s}(r)|^2 = 4 \left(\frac{Z}{a_0}\right)^3 e^{-2Zr/a_0} \frac{1}{4\pi}$$

where use has been made of the wave functions in Probs. 2.22 and 2.14. Go to dimensionless variables with

$$t \equiv \frac{2Zr}{a_0}$$

Then

$$V_C = \frac{2Ze^2}{a_0} \int d^3 t_1 \int d^3 t_2 \, \rho(t_1) \frac{e^2}{|\mathbf{t}_1 - \mathbf{t}_2|} \rho(t_2)$$

where now $\rho(t) \equiv e^{-t}/8\pi$. Just as in Eqs. (4.90)–(4.92), this becomes

$$V_C = \frac{Ze^2}{a_0} I$$

$$I \equiv \frac{1}{2} \int_0^\infty t_1^2 dt_1 \, e^{-t_1} \int_0^\infty t_2^2 dt_2 \, e^{-t_2} \left(\frac{1}{t_>}\right)$$

which is the expression in Eqs. (4.92) and (4.95).

The original expression can be written as

$$V_C = e \int d^3 r_1\, \rho(r_1) \Phi_C(r_1) = e \int 4\pi r_1^2 dr_1\, \rho(r_1) \Phi_C(r_1)$$

$$\Phi_C(r_1) = \int d^3 r_2\, \frac{e}{|\mathbf{r}_1 - \mathbf{r}_2|} \rho(r_2)$$

which is the interaction energy of the charge density of the first particle with the Coulomb potential created by the second.

Problem 4.8 As a model for the $(p\mu^- p)$ molecules ("mulecules") formed when μ^- are stopped in hydrogen, or for the $(pe^- p) \equiv H_2^+$ molecular ion, consider the problem of a particle of mass m at position $\mathbf{r}$ interacting through the Coulomb interaction with two heavy, fixed, point charges at positions $\mathbf{r}_1$ and $\mathbf{r}_2$. The hamiltonian for this problem is

$$H = \frac{\mathbf{p}^2}{2m} - \frac{e^2}{|\mathbf{r} - \mathbf{r}_1|} - \frac{e^2}{|\mathbf{r} - \mathbf{r}_2|} + \frac{e^2}{|\mathbf{r}_1 - \mathbf{r}_2|}$$

We shall compute the expectation value of this hamiltonian using a normalized linear combination of atomic wave functions with respect to each charge[3]

$$\psi(\mathbf{r}) = \mathcal{N} \left[\psi_{1s}(|\mathbf{r} - \mathbf{r}_1|) + \psi_{1s}(|\mathbf{r} - \mathbf{r}_2|) \right]$$

(a) Show the normalization constant is given by[4]

$$\frac{1}{\mathcal{N}^2} = 2 \left[1 + \left(1 + \Delta + \frac{1}{3}\Delta^2 \right) e^{-\Delta} \right] \qquad ; \ \Delta \equiv \frac{1}{a_0} |\mathbf{r}_2 - \mathbf{r}_1|$$

Here $\psi_{1s}(\mathbf{r}) = e^{-r/a_0}/(\pi a_0^3)^{1/2}$, $\varepsilon_{1s}^0 = -e^2/2a_0$, and $a_0 = \hbar^2/me^2$;

(b) Show the desired result can then be written

$$\text{B.E.} \equiv \frac{\langle \psi | H | \psi \rangle - \varepsilon_{1s}^0}{e^2/2a_0} = \frac{2}{\Delta} \left[\frac{(1 - 2\Delta^2/3)e^{-\Delta} + (1 + \Delta)e^{-2\Delta}}{1 + (1 + \Delta + \Delta^2/3)e^{-\Delta}} \right]$$

---

[3]This problem is longer, but the results are well worth it. The calculation provides the simplest description of the *molecular bond*. It uses the "linear combination of atomic orbitals (LCAO)" method.

[4]*Hint*: Establish the first result below, and use the second from the tables

$$\int d\Omega_t\, \frac{e^{-\alpha|\mathbf{t} - \mathbf{\Delta}|}}{|\mathbf{t} - \mathbf{\Delta}|} = \frac{2\pi}{\alpha x_> x_<} \left[e^{-\alpha(x_> - x_<)} - e^{-\alpha(x_> + x_<)} \right]$$

$$\int^x x^n e^{-ax}\, dx = -\frac{e^{-ax}}{a^{n+1}} [(ax)^n + n(ax)^{n-1} + \cdots + n!]$$

(c) Explain in what sense B.E. is an estimate of the binding energy of the molecule, and verify the plot of B.E.(Δ) in Fig. 4.1;

(d) How is the *exact* ground-state energy of this model problem related to what you have calculated?

(e) Discuss how you might improve this calculation, and this model;

(f) Compare with the experimental results for the H_2^+ ion (see Fig. 4.1)

$$r_e = 1.06\,\mathring{A} \qquad ; \text{ equilibrium internuclear separation}$$

$$\text{B.E.} = -2.65056\ \text{eV}$$

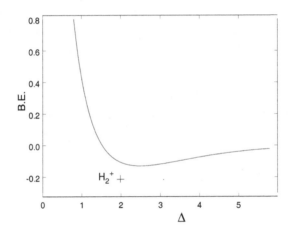

Fig. 4.1   The dimensionless quantity B.E.(Δ) in Prob. 4.8(c).

Solution to Problem 4.8

We will compute

$$E_0 = \frac{\langle \psi | H | \psi \rangle}{\langle \psi | \psi \rangle}$$

using the LCAO wave function for the lepton at position $\mathbf{r}$, where the two nuclei are located at the fixed positions $(\mathbf{r}_1, \mathbf{r}_2)$[5]

$$\psi(\mathbf{r}) = \mathcal{N}\left[\psi_{1s}(|\mathbf{r} - \mathbf{r}_1|) + \psi_{1s}(|\mathbf{r} - \mathbf{r}_2|)\right]$$

$$\psi_{1s}(\mathbf{r}) = \frac{1}{(\pi a_0^3)^{1/2}}\,e^{-r/a_0} \qquad\qquad ;\ \varepsilon_{1s}^0 = -\frac{e^2}{2a_0}$$

---

[5]Here $a_0 = \hbar^2/me^2$.

The variational principle tells us that if E is the true ground-state energy with the given hamiltonian

$$H = \frac{\mathbf{p}^2}{2m} - \frac{e^2}{|\mathbf{r} - \mathbf{r}_1|} - \frac{e^2}{|\mathbf{r} - \mathbf{r}_2|} + \frac{e^2}{|\mathbf{r}_1 - \mathbf{r}_2|}$$

then $E \le E_0$.[6]

As a preliminary to the calculation, we will need some integrals. First

$$\int \frac{e^{-\alpha|t-\mathbf{\Delta}|}}{|t - \mathbf{\Delta}|} d\Omega_t = \frac{2\pi}{\alpha x_> x_<} \left[e^{-\alpha(x_> - x_<)} - e^{-\alpha(x_> + x_<)} \right]$$

This is established with the following change of variable

$$|t - \mathbf{\Delta}| = \left(t^2 + \Delta^2 - 2t\Delta \cos\theta \right)^{1/2} \equiv u$$

$$du = \frac{-t\Delta}{(t^2 + \Delta^2 - 2t\Delta x)^{1/2}} dx \qquad ; \; x = \cos\theta$$

Then

$$\int \frac{e^{-\alpha|t-\mathbf{\Delta}|}}{|t - \mathbf{\Delta}|} d\Omega_t = \int_0^{2\pi} d\phi \int_{-1}^1 dx \frac{e^{-\alpha(t^2 + \Delta^2 - 2t\Delta x)^{1/2}}}{(t^2 + \Delta^2 - 2t\Delta x)^{1/2}}$$

$$= \frac{2\pi}{t\Delta} \int_{|t-\Delta|}^{t+\Delta} e^{-\alpha u} du$$

$$= \frac{2\pi}{t\Delta} \frac{1}{\alpha} \left[e^{-\alpha|t-\Delta|} - e^{-\alpha(t+\Delta)} \right]$$

which is the stated result, with $(x_>, x_<)$ the greater and lesser of (t, Δ).

Differentiation of this result with respect to α gives

$$\int e^{-\alpha|t-\mathbf{\Delta}|} d\Omega_t = \frac{2\pi}{\alpha^2 x_> x_<} \left\{ [1 + \alpha(x_> - x_<)] e^{-\alpha(x_> - x_<)} \right.$$

$$\left. - [1 + \alpha(x_> + x_<)] e^{-\alpha(x_> + x_<)} \right\}$$

---

[6]We know the exact solution to this problem in one limit. Write the separation of the two nuclei as $\mathbf{r}_2 - \mathbf{r}_1 \equiv a_0 \mathbf{\Delta}$. Then as $\Delta \to 0$, one has just the short-distance Coulomb repulsion of the two protons, and a lepton in a hydrogen-like atom with $Z = 2$

$$E \to \varepsilon_{1s}^0(Z = 2) + \frac{e^2}{\Delta a_0} \qquad ; \; \Delta \to 0$$

$$= 4\varepsilon_{1s}^0 + \frac{e^2}{\Delta a_0}$$

where $\varepsilon_{1s}^0 = -e^2/2a_0$. While this provides a nice check, we will not recover this limit from our variational calculation; we will be too high.

Two additional results follow from the integral tables

$$\int_0^\infty t^n e^{-t}\, dt = \Gamma(n+1) = n!$$

$$\int^x x^n e^{-ax}\, dx = \frac{-e^{-ax}}{a^{n+1}}\left[(ax)^n + n(ax)^{n-1} + \cdots + n!\right]$$

(a) The normalization is determined by the condition

$$\int d^3r\, \psi^\star \psi = N^2 \mathcal{I} = 1$$

where

$$\mathcal{I} = \int d^3r\, \left[\psi_{1s}^2(|\mathbf{r} - \mathbf{r}_1|) + \psi_{1s}^2(|\mathbf{r} - \mathbf{r}_2|) + 2\psi_{1s}(|\mathbf{r} - \mathbf{r}_1|)\psi_{1s}(|\mathbf{r} - \mathbf{r}_2|)\right]$$

$$\equiv \mathcal{I}_1 + \mathcal{I}_2 + \mathcal{I}_3$$

Consider the first integral. Change the integration variable to

$$\mathbf{r} - \mathbf{r}_1 \equiv \mathbf{t} \qquad ;\ d^3r = d^3t$$

The first integral $\mathcal{I}_1$ (as well as the second $\mathcal{I}_2$) is then just the normalization integral for hydrogen.

$$\mathcal{I}_1 = \mathcal{I}_2 = 1$$

In the third integral $\mathcal{I}_3$, introduce the dimensionless quantities

$$(\mathbf{r} - \mathbf{r}_1)/a_0 \equiv \boldsymbol{\rho} \qquad ;\ (\mathbf{r}_2 - \mathbf{r}_1)/a_0 \equiv \boldsymbol{\Delta}$$

Then

$$(\mathbf{r} - \mathbf{r}_2)/a_0 = \boldsymbol{\rho} - \boldsymbol{\Delta} \qquad ;\ d^3r/a_0^3 = d^3\rho$$

Written in terms of these quantities, $\mathcal{I}_3$ becomes

$$\mathcal{I}_3 = \frac{2}{\pi}\int e^{-\rho}\rho^2 d\rho d\Omega_\rho\, e^{-|\boldsymbol{\rho} - \boldsymbol{\Delta}|}$$

Introduce the previously derived relation for the angular integral to obtain

$$\mathcal{I}_3 = \frac{4}{\Delta}\int_0^\infty e^{-\rho}\rho^2 d\rho \frac{1}{\rho}$$
$$\times \left\{[1 + (x_> - x_<)]e^{-(x_> - x_<)} - [1 + (x_> + x_<)]e^{-(x_> + x_<)}\right\}$$
$$\equiv \frac{4}{\Delta}\left(\mathcal{I}_{3a} + \mathcal{I}_{3b} + \mathcal{I}_{3c} + \mathcal{I}_{3d}\right)$$

Note that the product $x_> x_< = \rho \Delta$ and the sum $x_> + x_< = \rho + \Delta$ are symmetric. The last two terms are therefore the simplest, and one finds

$$\mathcal{I}_{3c} = -\int_0^\infty e^{-\rho} \rho \, d\rho \, e^{-(\rho + \Delta)} = -e^{-\Delta} \int_0^\infty e^{-2\rho} \rho \, d\rho = -\frac{1}{4} e^{-\Delta}$$

$$\mathcal{I}_{3d} = -\int_0^\infty e^{-\rho} \rho \, d\rho \, (\rho + \Delta) e^{-(\rho + \Delta)}$$

$$= -e^{-\Delta} \left[\int_0^\infty e^{-2\rho} \rho^2 \, d\rho + \Delta \int_0^\infty e^{-2\rho} \rho \, d\rho \right]$$

$$= -\frac{1}{4} e^{-\Delta} (1 + \Delta)$$

For the first two terms, we separate the integration region on ρ into parts with $\Delta > \rho$ and $\Delta < \rho$. Then

$$\mathcal{I}_{3a} = \int_0^\Delta e^{-\rho} \rho \, d\rho \, e^{-(\Delta - \rho)} + \int_\Delta^\infty e^{-\rho} \rho \, d\rho \, e^{-(\rho - \Delta)}$$

$$= e^{-\Delta} \int_0^\Delta \rho \, d\rho + e^{\Delta} \int_\Delta^\infty e^{-2\rho} \rho \, d\rho$$

$$= e^{-\Delta} \frac{\Delta^2}{2} + e^{\Delta} \left[-\frac{e^{-2\rho}}{4} (2\rho + 1) \right]_\Delta^\infty$$

$$= e^{-\Delta} \left[\frac{\Delta^2}{2} + \frac{(2\Delta + 1)}{4} \right]$$

where use has been made of the previously given indefinite integral. In a similar manner

$$\mathcal{I}_{3b} = \int_0^\Delta e^{-\rho} \rho \, d\rho \, (\Delta - \rho) e^{-(\Delta - \rho)} + \int_\Delta^\infty e^{-\rho} \rho \, d\rho \, (\rho - \Delta) e^{-(\rho - \Delta)}$$

$$= e^{-\Delta} \int_0^\Delta \rho (\Delta - \rho) d\rho + e^{\Delta} \int_\Delta^\infty e^{-2\rho} \rho (\rho - \Delta) d\rho$$

$$= e^{-\Delta} \left[\frac{\Delta^3}{2} - \frac{\Delta^3}{3} \right] + e^{\Delta} \left[-\frac{e^{-2\rho}}{8} [(2\rho)^2 + 2(2\rho) + 2] + \Delta \frac{e^{-2\rho}}{4} (2\rho + 1) \right]_\Delta^\infty$$

$$= e^{-\Delta} \left[\frac{\Delta^3}{6} \right] + e^{-\Delta} \left[\frac{1}{8} (4\Delta^2 + 4\Delta + 2) - \frac{\Delta}{4} (2\Delta + 1) \right]$$

$$= e^{-\Delta} \left[\frac{\Delta^3}{6} + \frac{1}{4} (\Delta + 1) \right]$$

A combination of terms gives

$$
\mathcal{I} = 2 + e^{-\Delta}\frac{4}{\Delta}\left[-\frac{1}{4} - \frac{1}{4}(1+\Delta) + \frac{1}{2}\Delta^2 + \frac{1}{4}(2\Delta+1) + \frac{1}{6}\Delta^3 + \frac{1}{4}(\Delta+1)\right]
$$

$$
= 2\left[1 + e^{-\Delta}\left(1 + \Delta + \frac{1}{3}\Delta^2\right)\right]
$$

The normalization condition $N^2\mathcal{I} = 1$ is then as stated in the problem

$$
\frac{1}{N^2} = 2\left[1 + \left(1 + \Delta + \frac{1}{3}\Delta^2\right)e^{-\Delta}\right] \qquad ; \ \Delta \equiv \frac{1}{a_0}|\mathbf{r}_2 - \mathbf{r}_1|
$$

(b) We now compute $\langle\psi|H|\psi\rangle$ with this normalized wave function, making use of the one-body Schrödinger equations

$$
\left[\frac{\mathbf{p}^2}{2m} - \frac{e^2}{|\mathbf{r} - \mathbf{r}_1|}\right]\psi_{1s}(|\mathbf{r} - \mathbf{r}_1|) = \varepsilon_{1s}^0\psi_{1s}(|\mathbf{r} - \mathbf{r}_1|) \qquad ; \ \varepsilon_{1s}^0 = -\frac{e^2}{2a_0}
$$

$$
\left[\frac{\mathbf{p}^2}{2m} - \frac{e^2}{|\mathbf{r} - \mathbf{r}_2|}\right]\psi_{1s}(|\mathbf{r} - \mathbf{r}_2|) = \varepsilon_{1s}^0\psi_{1s}(|\mathbf{r} - \mathbf{r}_2|)
$$

The term $\varepsilon_{1s}^0 + e^2/\Delta a_0$ can then be factored from the integral, and the calculation is reduced to the evaluation of the following expression

$$
\langle\psi|H|\psi\rangle = \varepsilon_{1s}^0 + \frac{e^2}{\Delta a_0} + N^2\int d^3r\,[\psi_{1s}(|\mathbf{r} - \mathbf{r}_1|) + \psi_{1s}(|\mathbf{r} - \mathbf{r}_2|)]
$$

$$
\times\left[-\frac{e^2}{|\mathbf{r} - \mathbf{r}_1|}\psi_{1s}(|\mathbf{r} - \mathbf{r}_2|) - \frac{e^2}{|\mathbf{r} - \mathbf{r}_2|}\psi_{1s}(|\mathbf{r} - \mathbf{r}_1|)\right]
$$

In fact, the two remaining integrals are identical, since they differ only in the interchange $\mathbf{r}_1 \rightleftharpoons \mathbf{r}_2$; however, the whole integral is only a function of $|\mathbf{r}_2 - \mathbf{r}_1|$. Therefore

$$
\langle\psi|H|\psi\rangle = \varepsilon_{1s}^0 + \frac{e^2}{\Delta a_0} - 2e^2N^2\int d^3r
$$

$$
\times\left[\psi_{1s}(|\mathbf{r} - \mathbf{r}_2|)\frac{1}{|\mathbf{r} - \mathbf{r}_2|}\psi_{1s}(|\mathbf{r} - \mathbf{r}_1|) + \frac{1}{|\mathbf{r} - \mathbf{r}_2|}\psi_{1s}^2(|\mathbf{r} - \mathbf{r}_1|)\right]
$$

$$
\equiv \varepsilon_{1s}^0 + \frac{e^2}{\Delta a_0} - 2e^2N^2(\mathcal{J}_1 + \mathcal{J}_2)
$$

With the previous change of variables, the second integral becomes

$$
\mathcal{J}_2 \equiv \int d^3r\frac{1}{|\mathbf{r} - \mathbf{r}_2|}\psi_{1s}^2(|\mathbf{r} - \mathbf{r}_1|) = \frac{1}{\pi a_0}\int e^{-2\rho}\rho^2 d\rho\frac{1}{|\boldsymbol{\rho} - \boldsymbol{\Delta}|}d\Omega_\rho
$$

We note that this is just the Coulomb Green's function, and therefore

$$\mathcal{J}_2 = \frac{4}{a_0} \int e^{-2\rho} \rho^2 d\rho \left(\frac{1}{x_>} \right)$$

$$= \frac{4}{a_0} \left[\int_\Delta^\infty e^{-2\rho} \rho^2 d\rho \frac{1}{\rho} + \int_0^\Delta e^{-2\rho} \rho^2 d\rho \frac{1}{\Delta} \right]$$

$$= \frac{4}{a_0} \left\{ \left[-\frac{e^{-2\rho}}{4}(2\rho+1) \right]_\Delta^\infty + \left[-\frac{e^{-2\rho}}{8\Delta}[(2\rho)^2 + 2(2\rho) + 2] \right]_0^\Delta \right\}$$

$$= \frac{4}{a_0} \left[\frac{e^{-2\Delta}}{4}(2\Delta+1) - \frac{e^{-2\Delta}}{8\Delta}(4\Delta^2 + 4\Delta + 2) + \frac{1}{4\Delta} \right]$$

$$= \frac{1}{a_0} \left[\frac{1}{\Delta} - \frac{e^{-2\Delta}}{\Delta}(1+\Delta) \right]$$

A similar change in variables reduces the first integral to

$$\mathcal{J}_1 \equiv \int d^3r \, \psi_{1s}(|\mathbf{r}-\mathbf{r}_2|) \frac{1}{|\mathbf{r}-\mathbf{r}_2|} \psi_{1s}(|\mathbf{r}-\mathbf{r}_1|)$$

$$= \frac{1}{\pi a_0} \int e^{-\rho} \rho^2 d\rho \frac{e^{-|\boldsymbol{\rho}-\boldsymbol{\Delta}|}}{|\boldsymbol{\rho}-\boldsymbol{\Delta}|} d\Omega_\rho$$

The use of the previous result for the angular integral gives

$$\mathcal{J}_1 = \frac{2}{\Delta a_0} \int e^{-\rho} \rho^2 d\rho \frac{1}{\rho} \left[e^{-(x_> - x_<)} - e^{-(x_> + x_<)} \right]$$

$$\equiv \frac{2}{a_0 \Delta}(\tilde{\mathcal{J}}_{1a} + \tilde{\mathcal{J}}_{1b})$$

The second integral is

$$\tilde{\mathcal{J}}_{1b} = -\int_0^\infty e^{-\rho} \rho d\rho \, e^{-(\rho+\Delta)} = -e^{-\Delta} \int_0^\infty e^{-2\rho} \rho d\rho = -\frac{e^{-\Delta}}{4}$$

The first integral is

$$\tilde{\mathcal{J}}_{1a} = \int_0^\Delta e^{-\rho} \rho d\rho \, e^{-(\Delta-\rho)} + \int_\Delta^\infty e^{-\rho} \rho d\rho \, e^{-(\rho-\Delta)}$$

$$= e^{-\Delta} \int_0^\Delta \rho d\rho + e^{\Delta} \int_\Delta^\infty e^{-2\rho} \rho d\rho$$

$$= e^{-\Delta} \left(\frac{\Delta^2}{2} \right) + e^{\Delta} \left[-\frac{e^{-2\rho}}{4}(2\rho+1) \right]_\Delta^\infty$$

$$= e^{-\Delta} \left(\frac{\Delta^2}{2} \right) + e^{-\Delta} \frac{1}{4}(2\Delta+1)$$

A combination of terms gives

$$\mathcal{J}_1 = \frac{1}{a_0}e^{-\Delta}(1+\Delta)$$

The expectation value of H is therefore

$$\langle\psi|H|\psi\rangle = \varepsilon_{1s}^0 + \frac{e^2}{\Delta a_0} - 2e^2 N^2(\mathcal{J}_1 + \mathcal{J}_2)$$

$$= \varepsilon_{1s}^0 + \frac{e^2}{\Delta a_0} - \frac{2e^2 N^2}{a_0}\left[\frac{1}{\Delta} - \frac{1}{\Delta}(1+\Delta)e^{-2\Delta} + (1+\Delta)e^{-\Delta}\right]$$

If we form the combination

$$\text{B.E.} \equiv \frac{\langle\psi|H|\psi\rangle - \varepsilon_{1s}^0}{e^2/2a_0}$$

Then

$$\text{B.E.} = \frac{2}{\Delta} - \frac{2[1/\Delta - (1+\Delta)e^{-2\Delta}/\Delta + (1+\Delta)e^{-\Delta}]}{1 + (1+\Delta+\Delta^2/3)e^{-\Delta}}$$

$$= \frac{2}{\Delta}\left[\frac{1 + (1+\Delta+\Delta^2/3)e^{-\Delta} - 1 + (1+\Delta)e^{-2\Delta} - (\Delta+\Delta^2)e^{-\Delta}}{1 + (1+\Delta+\Delta^2/3)e^{-\Delta}}\right]$$

Hence we arrive at the given result

$$\text{B.E.} \equiv \frac{\langle\psi|H|\psi\rangle - \varepsilon_{1s}^0}{e^2/2a_0} = \frac{2}{\Delta}\left[\frac{(1-2\Delta^2/3)e^{-\Delta} + (1+\Delta)e^{-2\Delta}}{1 + (1+\Delta+\Delta^2/3)e^{-\Delta}}\right]$$

(c) The quantity B.E. provides a variational estimate of the difference in energy between the ground state of the molecular ion and a neutral atom plus free proton at rest; this is the minimum amount of energy required to dissociate the molecular ion. The dimensionless quantity B.E.(Δ) is plotted in Fig. 4.1. It has a minimum at which B.E.(Δ) is negative, indicating that the molecular ion is bound. The binding arises from the electrostatic attraction between the lepton and the two protons, the LCAO wave function providing an enhanced probability of finding the lepton between the nucleons.

(d) The variational principle states that the exact ground state of the starting hamiltonian lies below the curve in Fig. 4.1 at each internuclear separation Δ, which is simply a fixed parameter in the problem.

(e) This is a model hamiltonian, albeit a very good one, and there are many elements of physics left out. For example, missing are (with no attempt being made to estimate the size of the effects):

- Nuclear motion [more important for $(p\mu^- p)$ than $(pe^- p)$]; it is really a three-body problem;
- Relativistic corrections to the lepton motion;
- Spin-orbit interaction;
- Magnetic interaction between the lepton and nucleon moments;
- Finite size of the protons;
- Interaction with the quantized radiation field;
- Weak-neutral-current interactions, *etc.*

The calculation itself could be improved by using a more flexible variational wave function.[7]

(f) The experimental numbers for the H_2^+ molecular ion

$$r_e = 1.06 \text{ Å} \qquad ; \text{ equilibrium internuclear separation}$$
$$\text{B.E.} = -2.65056 \text{ eV}$$

are indicated in Fig. 4.1. The calculation provides a decent description of this molecule, and the experimental point indeed lies below the calculated curve.

****

(*Aside*) Although not required by the problem, we present the result of a simple variational calculation that improves the values of the internuclear separation and binding energy of the H_2^+ molecular ion. We use an ansatz for the wave function that depends upon an arbitrary parameter and then minimize the expectation value of the hamiltonian with respect to that parameter. We modify $\psi_{1s}(r)$ to

$$\psi_{1s}(r) = \left(\frac{\beta^3}{\pi a_0^3} \right)^{1/2} e^{-\beta r/a_0}$$

and again write the ansatz for the electron wave function as

$$\psi(\mathbf{r}) = \mathcal{N} \left[\psi_{1s}(|\mathbf{r} - \mathbf{r}_1|) + \psi_{1s}(|\mathbf{r} - \mathbf{r}_2|) \right]$$

Notice the presence of the arbitrary parameter β.

One can then repeat all the steps carried out in parts (a) and (b) to

---

[7]We should at least be able to reproduce the correct $\Delta \to 0$ limit of $E \to e^2/\Delta a_0 + 4\varepsilon_{1s}^0$ (see the previous footnote), or equivalently B.E.$(\Delta) \to 2/\Delta - 3$. This calculation gives B.E.$(\Delta) \to 2/\Delta - 2$, which indeed lies above the exact value.

finally obtain

$$\text{B.E.} = \frac{e^{-2\beta\Delta}\left\{\left[\beta\left(3\beta^2 - 10\beta + 3\right)\Delta^2 + 6\right]e^{\beta\Delta} + 6\beta\Delta + 6\right\}}{\Delta\left[e^{-\beta\Delta}\left(\beta^2\Delta^2 + 3\beta\Delta + 3\right) + 3\right]}$$

$$+ (\beta - 1)\frac{e^{-\beta\Delta}\left\{\beta\left[3 - \beta(\beta+1)\Delta^2\right] - 3\right\} + 3(\beta - 1)}{e^{-\beta\Delta}\left(\beta^2\Delta^2 + 3\beta\Delta + 3\right) + 3}$$

Now B.E. is a function of β. For $\beta = 1$, the second contribution vanishes and this expression reduces to the one calculated before.

The minimum of the B.E. occurs for

$$\beta = 1.238 \qquad ; \text{ minimizes B.E.}$$

In fact, the variational principle allows one to minimize B.E.(Δ) with respect to β at each value of Δ, which simply appears as an external parameter in the calculation. To a very good approximation, these values of β are given by

$$\beta = 1 + 1.765\, e^{-\Delta} \qquad ; \text{ minimum at each } \Delta$$

The absolute minimum then occurs at $\Delta = 2.003$, which gives the above value of $\beta = 1.238$

Figure 4.2 shows a comparison between this new variational calculation and the previous results for the H_2^+ molecular ion in Fig. 4.1.

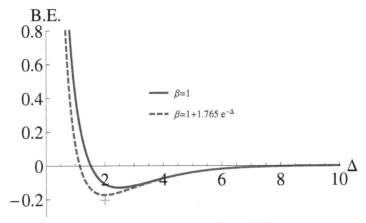

Fig. 4.2   The dimensionless quantity B.E.(Δ) in Prob. 4.8(c) (solid curve, corresponding to $\beta = 1$) and the variationally improved result obtained for $\beta = 1 + 1.7646\, e^{-\Delta}$ (dashed curve).

The previous minimum (with $\beta = 1$) occurred at

$$r_e = 1.32 \,\overset{\circ}{A} \qquad ; \beta = 1.000$$
$$\text{B.E.} = -1.7634 \text{ eV}$$

The new minimum (with $\beta = 1.238$) occurs at

$$r_e = 1.06 \,\overset{\circ}{A} \qquad ; \beta = 1.238$$
$$\text{B.E.} = -2.3523 \text{ eV}$$

The experimental numbers for the H_2^+ molecular ion are

$$r_e = 1.06 \,\overset{\circ}{A} \qquad ; \text{experiment}$$
$$\text{B.E.} = -2.65056 \text{ eV}$$

Thus we see that the use of a variational parameter leads to a fine determination of the separation distance of the protons and to a much more precise value for the binding energy.

****

Problem 4.9 (a) Show the electromagnetic fields $(\mathbf{B}, \mathbf{E})$ are unchanged under the following gauge transformation of the potentials in Eqs. (4.130)

$$\mathbf{A} \to \mathbf{A} + \boldsymbol{\nabla}\Lambda \qquad ; \Phi \to \Phi - \frac{1}{c}\frac{\partial \Lambda}{\partial t}$$

where $\Lambda(\mathbf{r}, t)$ is a scalar function of position and time;

(b) Show that the time-dependent Schrödinger equation is left *invariant* under the gauge transformation in part (a) accompanied by the following *local phase transformation* of the Schrödinger wave function

$$\Psi(\mathbf{r}, t) \to e^{ie\Lambda(\mathbf{r},t)/\hbar c}\Psi(\mathbf{r}, t)$$

Solution to Problem 4.9

(a) The fields are related to the potentials through Eqs. (4.130)

$$\mathbf{B} = \boldsymbol{\nabla} \times \mathbf{A}$$
$$\mathbf{E} = -\boldsymbol{\nabla}\Phi - \frac{1}{c}\frac{\partial \mathbf{A}}{\partial t}$$

Substitution of the gauge-transformed potentials gives

$$\mathbf{B} \to \nabla \times (\mathbf{A} + \nabla \Lambda) = \nabla \times \mathbf{A}$$

$$\mathbf{E} \to -\nabla \left(\Phi - \frac{1}{c} \frac{\partial \Lambda}{\partial t} \right) - \frac{1}{c} \frac{\partial}{\partial t} (\mathbf{A} + \nabla \Lambda) = -\nabla \Phi - \frac{1}{c} \frac{\partial \mathbf{A}}{\partial t}$$

Hence the physical fields $(\mathbf{E}, \mathbf{B})$ are unchanged under a gauge transformation.

(b) Consider the time-dependent Schrödinger equation arising from Eq. (4.129)

$$i\hbar \frac{\partial \Psi}{\partial t} = \left\{ \frac{1}{2m_0} \left[\mathbf{p} - \frac{e}{c} \mathbf{A}(\mathbf{r}, t) \right]^2 + V(r) + e\Phi(\mathbf{r}, t) \right\} \Psi$$

Now simultaneously make the gauge transformation and the indicated local change in the phase of the wave function. The l.h.s. of the Schrödinger equation is modified according to

$$\text{l.h.s.} \to e^{ie\Lambda/\hbar c} \left(i\hbar \frac{\partial}{\partial t} - \frac{e}{c} \frac{\partial \Lambda}{\partial t} \right) \Psi$$

With $\mathbf{p} = (\hbar/i)\nabla$, the r.h.s. of the Schrödinger equation becomes

$$\text{r.h.s.} \to e^{ie\Lambda/\hbar c} \left\{ \frac{1}{2m_0} \left[\frac{\hbar}{i} \nabla + \frac{e}{c} \nabla \Lambda - \frac{e}{c} (\mathbf{A} + \nabla \Lambda) \right]^2 + V + e\Phi - \frac{e}{c} \frac{\partial \Lambda}{\partial t} \right\} \Psi$$

A cancellation of terms, and of the overall phase, then reproduces the starting Schrödinger equation.

Problem 4.10 Consider a particle of mass m in a one-dimensional box of size L with a potential $2mV/\hbar^2 = v\delta(x)$ located at the center (see Fig. 4.3).

(a) Calculate the energy functional $\mathcal{E}(\alpha)$ using the trial wave function

$$\psi(x) = \mathcal{N} \sin \alpha(x + L/2) \qquad ; \ -L/2 \leq x \leq 0$$
$$= \mathcal{N} \sin \alpha(L/2 - x) \qquad ; \ 0 \leq x \leq L/2$$

(b) Obtain the first-order perturbation theory result using $\alpha = \pi/L$;

(c) Minimize $\mathcal{E}(\alpha)$ with respect to α to find an improved variational estimate;

(d) Compare with the exact answer obtained by matching boundary conditions across the potential [see Eqs. (2.320)–(2.323)].

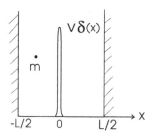

Fig. 4.3 Particle of mass m in a one-dimensional box of size L with an additional potential $2mV/\hbar^2 = v\delta(x)$ in the center.

Solution to Problem 4.10

(a) It is straightforward to evaluate the normalization constant $\mathcal{N}$. Use

$$\int_{-L/2}^{0} dx \, \sin^2\left[\alpha\left(x + \frac{L}{2}\right)\right] = \int_{0}^{L/2} dx \, \sin^2\left[\alpha\left(\frac{L}{2} - x\right)\right]$$

$$= \frac{1}{4\alpha}[\alpha L - \sin(\alpha L)]$$

Thus $\mathcal{N}^2$ is given by

$$\mathcal{N}^2 = \frac{2\alpha}{\alpha L - \sin(\alpha L)}$$

We can then express $\psi(x)$ in the box as

$$\psi(x) = \mathcal{N}\left[\theta(-x)\sin\alpha\left(x + \frac{L}{2}\right) + \theta(x)\sin\alpha\left(\frac{L}{2} - x\right)\right]$$

$$\psi(0) = \mathcal{N}\sin\frac{\alpha L}{2}$$

where $\theta(x)$ is the step function. We use this analytic expression for the wave function in the box because in computing the kinetic energy, we have to be concerned with a possible contribution coming from the discontinuity in the derivative at $x = 0$. We will use the following properties of the step function $\theta(x)$ and delta function $\delta(x)$

$$\theta(x) = \int^{x} dx \, \delta(x)$$

$$\frac{d}{dx}\theta(x) = \delta(x) = -\frac{d}{dx}\theta(-x)$$

Furthermore, with the aid of a partial integration, one has for $x > 0$

$$\int^x dx\, f(x) \frac{d}{dx} \delta(x) = -f'(0) \qquad ; x > 0$$

Around $x = 0$, the ansatz behaves as

$$\psi(x) = \mathcal{N} \left\{ \sin \frac{\alpha L}{2} + [\theta(-x) - \theta(x)]\, \alpha x \cos \left(\frac{\alpha L}{2} \right) - \frac{\alpha^2 x^2}{2} \sin \left(\frac{\alpha L}{2} \right) + \cdots \right\}$$

The singular part of the second derivative of $\psi(x)$ in this region is then calculated to be

$$\frac{1}{\mathcal{N}} \left. \frac{d^2 \psi(x)}{dx^2} \right|_{\text{sing.}} = -4\alpha \delta(x) \cos \left(\frac{\alpha L}{2} \right) - 2\alpha x\, \delta'(x) \cos \left(\frac{\alpha L}{2} \right)$$

This leads to an additional contribution to the kinetic energy coming from the discontinuity in slope of

$$-\frac{\hbar^2}{2m} \int dx \left[\psi^*(x) \frac{d^2}{dx^2} \psi(x) \right]_{\text{sing.}} = \frac{\hbar^2 \alpha^2}{2m} \mathcal{N}^2 \frac{\sin \alpha L}{\alpha}$$

We are now in position to evaluate the energy functional in Eq. (4.2)

$$\mathcal{E}(\alpha) = \frac{\hbar^2 \alpha^2}{2m} \mathcal{N}^2 \left[\int_{-L/2}^{0} dx\, \sin^2 \alpha \left(x + \frac{L}{2} \right) + \int_{0}^{L/2} dx\, \sin^2 \alpha \left(\frac{L}{2} - x \right) \right.$$
$$\left. + \frac{\sin \alpha L}{\alpha} \right] + \frac{\hbar^2 v}{2m} \mathcal{N}^2 \sin^2 \frac{\alpha L}{2}$$

where we have included the potential $V(x) = (\hbar^2 v / 2m)\delta(x)$. With the use of the previous integrals, this gives

$$\mathcal{E}(\alpha) = \frac{\hbar^2 \mathcal{N}^2}{2m} \left[\frac{\alpha^2 L}{2} + \frac{\alpha}{2} \sin \alpha L + v \sin^2 \frac{\alpha L}{2} \right]$$

Hence the energy functional is given by

$$\mathcal{E}(\alpha) = \frac{\hbar^2}{2mL^2} \frac{\alpha L \left[(\alpha L)^2 + \alpha L \sin (\alpha L) - vL \cos (\alpha L) + vL \right]}{\alpha L - \sin (\alpha L)}$$

(b) It is straightforward to evaluate the energy functional at $\alpha = \pi/L$; one obtains

$$\mathcal{E} \left(\frac{\pi}{L} \right) = \frac{\hbar^2 \pi^2}{2mL^2} + \frac{\hbar^2 v}{mL}$$

The first term is the ground-state energy of a particle in a one-dimensional box of size L, whose eigenfunctions in the box are

$$\phi_n(x) = \sqrt{\frac{2}{L}} \sin\left[\frac{\pi n\,(x + L/2)}{L}\right] \qquad ; n = 1, 2, \cdots$$

The first-order perturbation theory result for the ground state is easily obtained from $V(x) = (\hbar^2 v/2m)\delta(x)$ as

$$E^{(1)} = \int_{-L/2}^{L/2} \phi_1^\star(x) V(x) \phi_1(x)\,dx = \frac{\hbar^2 v}{mL}$$

This corresponds to the second term appearing in the expression for $\mathcal{E}(\pi/L)$.

(c) Consider the dimensionless quantity

$$F(\alpha L) \equiv \frac{2mL^2}{\hbar^2}\mathcal{E}(\alpha)$$

and rename $\beta \equiv \alpha L$. Then, after some algebra,

$$\frac{dF}{d\beta} = \frac{\left(2\beta^2 - 1\right)\cos\beta/2 - 2\beta\sin\beta/2 + \cos 3\beta/2}{(\beta - \sin\beta)^2}\left(vL\sin\beta/2 + 2\beta\cos\beta/2\right)$$

An extremum of F is obtained if

$$\tan\frac{\beta}{2} = -\frac{2\beta}{vL} \qquad ; \beta = \alpha L$$

It is relatively easy to convince oneself graphically that this corresponds to a minimum of $F(\alpha L)$ [see Fig. 4.4].

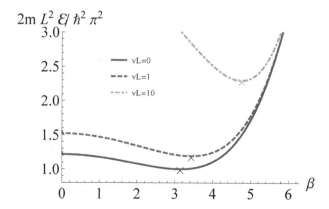

Fig. 4.4   The dimensionless quantity $2mL^2\mathcal{E}/\hbar^2\pi^2$ as a function of $\beta = \alpha L$, for different values of vL. The minimum for each vL is indicated with a cross.

(d) The exact solution is found by matching boundary conditions about the potential: the form of $\psi(x)$ given in part (a) may be an exact solution to the Schrödinger equation if both Eqs. (2.320) and (2.323) are fulfilled. While the first equation is automatically satisfied by $\psi(x)$, the second equation reduces to

$$-v \sin\left(\frac{L\alpha}{2}\right) - 2\alpha \cos\left(\frac{L\alpha}{2}\right) = 0$$

which is precisely the equation obtained in part (c)

$$\tan\frac{\beta}{2} = -\frac{2\beta}{vL} \qquad ; \beta = \alpha L$$

We thus conclude that in this case the variational theorem provides the exact answer.

****

(*Aside*) We may find an analytic approximation to the energy using the result in part (b) looking for a solution of the form

$$\tilde{\beta} = c_0 + c_1 vL + c_2(vL)^2 + \cdots$$

If one substitutes this expression inside the equation which determines the minimum and then expands in powers of vL, then one finds an infinite sequence of equations, corresponding to given powers of vL, which can be solved sequentially, starting from the lowest power of vL. This procedure allows to find a large number of coefficients; for example, one finds

$$c_0 = \pi \qquad\qquad ; c_1 = \frac{1}{\pi}$$

$$c_2 = -\frac{1}{\pi^3} \qquad\qquad ; c_3 = \frac{2}{\pi^5} - \frac{1}{12\pi^3}$$

$$c_4 = \frac{1}{3\pi^5} - \frac{5}{\pi^7}$$

$$c_5 = \frac{14}{\pi^9} - \frac{5}{4\pi^7} + \frac{1}{80\pi^5} \qquad ; etc.$$

Therefore the improved variational estimate is obtained evaluating $F(\tilde{\beta})$

$$\mathcal{E}_{var} = \frac{\hbar^2}{2mL^2} F(\tilde{\beta}) = \frac{\hbar^2}{2mL^2}\left[\pi^2 + 2(vL) - \frac{(vL)^2}{\pi^2} + \frac{(vL)^3\left(12 - \pi^2\right)}{6\pi^4} + \cdots\right]$$

This result is compared with the exact answer in Fig. 4.5.

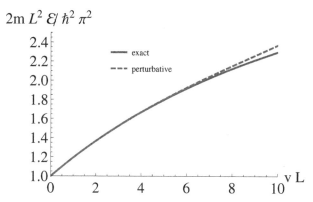

$2m\,L^2\,\mathcal{E}/\hbar^2\,\pi^2$

Fig. 4.5 The dimensionless quantity $\mathcal{E}_{\mathrm{var}}/(\hbar^2\pi^2/2mL^2)$ as a function of vL. The solid line is the exact result, while the dashed line is the perturbative expression calculated above.

****

Problem 4.11 Verify the behavior of the spherical harmonics under spatial reflection in Eq. (4.184) from their defining relation in Eq. (2.172).

Solution to Problem 4.11

The Legendre and associated Legendre polynomials are defined in Eqs. (2.170)

$$\Phi_m(\phi) = \frac{1}{\sqrt{2\pi}}e^{im\phi} \qquad\qquad ;\ m = 0, \pm1, \pm2, \cdots$$

$$P_l^m(x) \equiv (1 - x^2)^{|m|/2}\frac{d^{|m|}}{dx^{|m|}}P_l(x)$$

$$P_l(x) = \frac{1}{2^l\,l!}\frac{d^l}{dx^l}(x^2 - 1)^l \qquad ;\ l = 0, 1, 2, \cdots$$

where $x = \cos\theta$. The spherical harmonics are then defined in Eqs. (2.172)

$$Y_{lm}(\theta, \phi) \equiv (-1)^m\left[\frac{2l+1}{4\pi}\frac{(l-m)!}{(l+m)!}\right]^{1/2}P_l^m(\cos\theta)e^{im\phi} \quad ;\ m \geq 0$$

$$Y_{lm}^\star \equiv (-1)^m Y_{l,-m} \qquad\qquad\qquad ;\ \text{defines others}$$

We are to determine the behavior of the spherical harmonics under the transformation $(\theta, \phi) \to (\pi - \theta, \phi + \pi)$. Since $\cos(\pi - \theta) = -\cos\theta$, this is the same as

$$(x, \phi) \to (-x, \phi + \pi)$$

Under the transformation $x \to -x$, the Legendre and associated Legendre polynomials behave as

$$P_l(-x) = (-1)^l P_l(x)$$
$$P_l^m(-x) = (-1)^{l+|m|} P_l^m(x)$$

The spherical harmonics contain the factor $e^{im\phi}$ for all values of the integer m, and

$$e^{im(\phi+\pi)} = \left(e^{i\pi}\right)^m e^{im\phi} = (-1)^m e^{im\phi}$$

Hence the spherical harmonics transform as

$$Y_{lm}(\pi - \theta, \phi + \pi) = (-1)^{l+|m|+m} Y_{lm}(\theta, \phi)$$
$$= (-1)^l Y_{lm}(\theta, \phi)$$

This is Eq. (4.184).

Problem 4.12 Make a good numerical calculation and three-dimensional plot[8] of the charge density in the eigenstates $\chi_\pm(\mathbf{r})$ in Eqs. (4.200) (see Figs. 4.17–4.18 in the text.)

Solution to Problem 4.12

The shifted $n = 2$ eigenstates $\chi_\pm(\mathbf{r})$ in the linear Stark effect are given in Eqs. (4.200)

$$\chi_\pm(\mathbf{r}) = \frac{1}{\sqrt{2}} \left[\psi_{200}(\mathbf{r}) \pm \psi_{210}(\mathbf{r}) \right]$$

where the hydrogen-like wave functions are

$$\psi_{200}(r, \theta, \phi) = \left(\frac{Z}{2a_0} \right)^{3/2} (2 - \rho) \, e^{-\rho/2} \, Y_{00}(\theta, \phi)$$

$$\psi_{21m}(r, \theta, \phi) = \frac{1}{\sqrt{3}} \left(\frac{Z}{2a_0} \right)^{3/2} \rho \, e^{-\rho/2} \, Y_{1m}(\theta, \phi)$$

and $\rho(r) = Zr/a_0$.

---

[8]Your choice as to just how to do this.

In Fig. 4.6 we plot the quantities $P_\pm(\theta, \phi)$, with

$$P_\pm(\theta, \phi) \equiv \frac{a_0^3}{Z^3} |\chi_\pm(a_0/Z, \theta, \phi)|^2$$

$$= \frac{1}{64\pi} e^{-1} |1 \pm \cos(\theta)|^2$$

These are directly related to the probability densities. Notice how these are displaced from the origin, leading to a dipole moment for the states.[9]

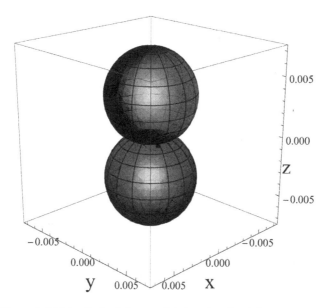

Fig. 4.6   The probabilities $P_+(\theta, \phi)$ (upper globe) and $P_-(\theta, \phi)$ (lower globe) at $\rho(r) = 1$.

Problem 4.13 A particle moves in a two-dimensional circular box of radius a with perfectly rigid walls. Use the radial wave function $R(r) = [1 - (r/a)^\alpha]$, with α as a variational parameter, to calculate the ground-state energy. Compare with the exact answer obtained from the eigenvalue equation $J_0(ka) = 0$ (see Prob. 2.30).

Solution to Problem 4.13

With the use of the radial wave function $R(r)$ given in the problem, we

---

[9]Compare Fig. 4.18 in the text; it is the integral over all ρ that reverses the sign of the effect [see Eq. (4.190)].

may evaluate the expectation value of the hamiltonian of a free particle in two dimensions.

$$E_{\text{var}}(\alpha) = \frac{\int_0^a r \, dr \, R(r) H R(r)}{\int_0^a r \, dr \, [R(r)]^2}$$

$$H = -\frac{\hbar^2}{2m} \nabla^2 \doteq -\frac{\hbar^2}{2m} \frac{1}{r} \frac{d}{dr} r \frac{d}{dr}$$

The denominator in $E_{\text{var}}(\alpha)$ is

$$\int_0^a r \, dr \, [R(r)]^2 = a^2 \int_0^1 t \, dt \, \left(1 - 2t^\alpha + t^{2\alpha}\right)$$

$$= a^2 \left[\frac{1}{2} - \frac{2}{\alpha + 2} + \frac{1}{2(\alpha + 1)}\right]$$

$$= \frac{a^2}{2(\alpha + 1)(\alpha + 2)} \left[(\alpha + 1)(\alpha + 2) - 4(\alpha + 1) + (\alpha + 2)\right]$$

$$= \frac{a^2 \alpha^2}{2(\alpha + 1)(\alpha + 2)}$$

The numerator in $E_{\text{var}}(\alpha)$ is

$$\int_0^a r \, dr \, R(r) H R(r) = \frac{\hbar^2}{2m} \alpha \int_0^1 dt \, (1 - t^\alpha) \frac{d}{dt} t^\alpha$$

$$= \frac{\hbar^2}{2m} \alpha^2 \int_0^1 dt \, \left(t^{\alpha - 1} - t^{2\alpha - 1}\right)$$

$$= \frac{\hbar^2}{2m} \alpha^2 \left(\frac{1}{\alpha} - \frac{1}{2\alpha}\right) = \frac{\hbar^2}{2m} \frac{\alpha}{2}$$

Hence the result for the variational estimate is

$$E_{\text{var}}(\alpha) = \frac{\hbar^2}{2a^2 m} \left(\alpha + 3 + \frac{2}{\alpha}\right)$$

Since the variational theorem implies that $E_{\text{var}} \geq E_{\text{exact}}$, the optimal estimate of the energy is obtained by minimizing the above expression with respect to α

$$\frac{d}{d\alpha} E_{\text{var}}(\alpha) = \frac{\hbar^2}{2a^2 m} \left(1 - \frac{2}{\alpha^2}\right) = 0$$

$$\implies \qquad \alpha = \sqrt{2}$$

Thus E_{var} has a minimum for $\alpha = \sqrt{2}$, and for this value[10]

$$E_{\text{var}}(\sqrt{2}) = \frac{\hbar^2}{2a^2 m}\left(3 + 2\sqrt{2}\right) = 5.828\,\frac{\hbar^2}{2a^2 m}$$

The exact energy for a particle confined in a circular box of radius a has been calculated in Prob. 2.30, and it reads

$$E_{\text{exact}} = \frac{\hbar^2}{2a^2 m}Z_{10}^2 = 5.783\,\frac{\hbar^2}{2a^2 m}$$

The dimensionless quantity $2ma^2 E/\hbar^2$ is plotted for the two cases in Fig. 4.7. Notice that $E_{\text{var}}(\sqrt{2})$ is precise to less than 1%.

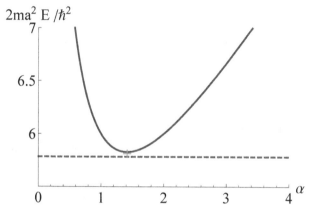

Fig. 4.7   The dimensionless quantity $2ma^2 E_{\text{var}}/\hbar^2$ plotted as function of α (solid line); the dashed horizontal line is the exact result.

Problem 4.14 Expand the wave function $\psi(x)$ in a truncated basis of known wave functions $\phi_n(x)$ satisfying the appropriate boundary conditions

$$\psi(x) = \sum_{n=1}^{N} a_n \phi_n(x)$$

and compute the energy functional

$$\mathcal{E}(\psi) = \frac{\langle \psi|H|\psi\rangle}{\langle \psi|\psi\rangle}$$

---

[10]One must restrict the parameter space to $\alpha > 0$, since for $\alpha < 0$ the variational ansatz would be singular at the origin; here we quote numbers to four significant figures.

Treat the *coefficients* $(a_1, a_2, \cdots, a_N)$ as variational parameters, and show that the Euler-Lagrange equations for this variational problem are equivalent to diagonalizing the matrix H_{mn} in the truncated basis. Hence conclude that this analysis provides a variational basis for *matrix mechanics*.

Solution to Problem 4.14

Substitute the expansion of the wave function into the energy functional

$$\mathcal{E} = \frac{\sum_n \sum_m a_m^\star a_n \langle \phi_m | H | \phi_n \rangle}{\sum_n a_n^\star a_n}$$

Now compute the change of the energy functional when the coefficients are varied by

$$a_n \to a_n + \delta a_n \qquad ; \; a_n^\star \to a_n^\star + \delta a_n^\star$$

The expansion of the denominator in the energy functional makes the following contribution to this change

$$\frac{\sum_n \sum_m a_m^\star a_n \langle \phi_m | H | \phi_n \rangle}{\sum_n (a_n^\star a_n + \delta a_n^\star a_n + a_n^\star \delta a_n)} \approx \mathcal{E} \left(1 - \frac{\sum_n (\delta a_n^\star a_n + a_n^\star \delta a_n)}{\sum_n |a_n|^2} \right)$$

The variational principle in Eq. (4.20) implies that the first variation of the energy functional is to be set equal to zero. After multiplication by $\sum_n |a_n|^2$, the result is

$$\sum_n \delta a_n \sum_m a_m^\star \left(\langle \phi_m | H | \phi_n \rangle - \mathcal{E} \delta_{mn} \right)$$

$$+ \sum_m \delta a_m^\star \sum_n \left(\langle \phi_m | H | \phi_n \rangle - \mathcal{E} \delta_{mn} \right) a_n = 0$$

where the terms in $\mathcal{E}$ come from the expansion of the denominator. This expression is to vanish for arbitrary, independent, variations $(\delta a_n, \delta a_n^\star)$. One then has from the coefficient of $\delta a_m^\star$

$$\sum_{n=1}^N \left(\langle \phi_m | H | \phi_n \rangle - \mathcal{E} \delta_{mn} \right) a_n = 0 \qquad ; \; m = 1, 2, \cdots, N$$

This is the set of equations of matrix mechanics. From the coefficient of δa_n one concludes

$$\sum_{m=1}^N a_m^\star \left(\langle \phi_m | H | \phi_n \rangle - \mathcal{E} \delta_{mn} \right) = 0 \qquad ; \; n = 1, 2, \cdots, N$$

It follows from the hermiticity of H that this set of equations is just the complex conjugate of the first set.

Problem 4.15 The relativistic kinetic energy of an electron is given by

$$T = \left(\mathbf{p}^2 c^2 + m_0^2 c^4\right)^{1/2} - m_0 c^2 = \frac{\mathbf{p}^2}{2m_0} - \frac{\mathbf{p}^4}{8m_0^3 c^2} + \cdots$$

Treat the second term on the right as a perturbation, and calculate the first-order shift in the ground-state energy of hydrogen-like atoms (with charge Z on the nucleus). Compare with the unperturbed value.

Solution to Problem 4.15

The first-order energy shift is given by

$$\delta E_{1s}^{\text{rel}} = \langle \phi_{1s}|H'|\phi_{1s} \rangle$$

where the perturbation is

$$H' = -\frac{\mathbf{p}^4}{8m_0^3 c^2}$$

From Prob. 2.22, the $1s$ wave function is

$$\phi_{1s}(r) = \left(\frac{Z}{a_0}\right)^{3/2} 2\,e^{-Zr/a_0} Y_{00}$$

With the observation that $\mathbf{p}^2$ is hermitian, the energy shift can then be written as

$$\delta E_{1s}^{\text{rel}} = -\frac{4}{8m_0^3 c^2}\left(\frac{Z}{a_0}\right)^3 \int r^2 dr \left|\mathbf{p}^2 e^{-Zr/a_0}\right|^2$$

The radial part of the laplacian is given in Eq. (2.200) as

$$\mathbf{p}^2 = -\hbar^2 \nabla^2 \doteq -\frac{\hbar^2}{r}\frac{\partial^2}{\partial r^2} r$$

Therefore

$$\delta E_{1s}^{\text{rel}} = -\frac{4\hbar^4}{8m_0^3 c^2}\left(\frac{Z}{a_0}\right)^3 \int dr \left|\frac{d^2}{dr^2}\left(r e^{-Zr/a_0}\right)\right|^2$$

$$= -\frac{4\hbar^4}{8m_0^3 c^2}\left(\frac{Z}{a_0}\right)^4 \int du \left|\frac{d^2}{du^2}\left(u e^{-u}\right)\right|^2 \qquad ; u \equiv \frac{Zr}{a_0}$$

The constants can be combined using $a_0 = \hbar/\alpha m_0 c$, and the derivatives taken, to obtain

$$\delta E_{1s}^{\text{rel}} = -\frac{1}{2}\left(Z^4 \alpha^4\right) m_0 c^2\, \mathcal{I}$$

$$\mathcal{I} \equiv \int_0^\infty du\,(u-2)^2\, e^{-2u}$$

With $2u \equiv t$, the integral is evaluated as[11]

$$\mathcal{I} = \frac{1}{2}\int_0^\infty dt\,\left(\frac{t^2}{4} - 2t + 4\right) e^{-t} = \frac{1}{2}\left(\frac{1}{2} - 2 + 4\right) = \frac{5}{4}$$

Hence

$$\delta E_{1s}^{\text{rel}} = -\frac{5}{8}\left(Z^4 \alpha^4\right) m_0 c^2$$

From Eq. (2.266), the unperturbed energy of the $1s$ state is

$$E_{1s}^0 = -\frac{1}{2}(Z^2 \alpha^2) m_0 c^2$$

The fractional energy shift from the relativistic correction is therefore[12]

$$\frac{\delta E_{1s}^{\text{rel}}}{E_{1s}^0} = \frac{5}{4}(Z^2 \alpha^2)$$

Problem 4.16[13] A one-dimensional harmonic oscillator is perturbed by an extra potential energy γx^3. Calculate the change in each energy level to second order in the perturbation. Discuss.

Solution to Problem 4.16

With a perturbation $H' = \gamma x^3$, the diagonal matrix element vanishes, and the second-order energy shift of the oscillator levels is given by Eq. (4.60) as

$$\delta E_n^{(2)} = \gamma^2 \sum_{m \neq n} \frac{|\langle m|x^3|n\rangle|^2}{E_n^0 - E_m^0}$$

---

[11] Use $\int_0^\infty t^n e^{-t}\, dt = n!$.

[12] This result is identical to that given in [Schiff (1968)] for the $(Z^2 \alpha^2)$ correction to E_{1s}^0 obtained by solving a relativistic Schrödinger equation in a point Coulomb potential.

[13] See problem 8.2 in [Schiff (1968)].

From Eqs. (3.161), the coordinate x is written in terms of the creation and destruction operators as[14]

$$x = \left(\frac{\hbar}{2m_0\omega_0}\right)^{1/2}(a^\dagger + a)$$

Hence

$$\delta E_n^{(2)} = \gamma^2 \left(\frac{\hbar}{2m_0\omega_0}\right)^3 \sum_{m\neq n} \frac{|\langle m|(a^\dagger + a)^3|n\rangle|^2}{E_n^0 - E_m^0}$$

There are four states contributing to the sum, $|m\rangle = |n \pm 3\rangle$ and $|m\rangle = |n \pm 1\rangle$. The required matrix elements are computed from the properties of the creation and destruction operators

$$a^\dagger|n\rangle = \sqrt{n+1}\,|n+1\rangle \qquad ; \; a|n\rangle = \sqrt{n}\,|n-1\rangle$$

The contributions to the sum from the first pair of states are

$$\frac{|\langle n+3|a^\dagger a^\dagger a^\dagger|n\rangle|^2}{-3\hbar\omega_0} = \frac{-1}{3\hbar\omega_0}[(n+3)(n+2)(n+1)]$$

$$\frac{|\langle n-3|aaa|n\rangle|^2}{3\hbar\omega_0} = \frac{1}{3\hbar\omega_0}[(n-2)(n-1)n]$$

The contributions to the sum from the second pair of states are

$$\frac{|\langle n+1|(a^\dagger + a)^3|n\rangle|^2}{-\hbar\omega_0} = \frac{-1}{\hbar\omega_0}\left|\langle n+1|a^\dagger a^\dagger a + a^\dagger a a^\dagger + a a^\dagger a^\dagger|n\rangle\right|^2$$

$$= \frac{-1}{\hbar\omega_0}\left|(n + n+1 + n+2)\sqrt{n+1}\right|^2$$

$$= \frac{-9}{\hbar\omega_0}(n+1)^3$$

and

$$\frac{|\langle n-1|(a^\dagger + a)^3|n\rangle|^2}{\hbar\omega_0} = \frac{1}{\hbar\omega_0}\left|\langle n-1|a^\dagger aa + aa^\dagger a + aaa^\dagger|n\rangle\right|^2$$

$$= \frac{1}{\hbar\omega_0}\left|(n-1 + n + n+1)\sqrt{n}\right|^2$$

$$= \frac{9}{\hbar\omega_0}n^3$$

---

[14] For ease of writing, we here suppress the hats on the operators.

A combination of these results gives

$$\delta E_n^{(2)} = \frac{\gamma^2}{\hbar\omega_0} \left(\frac{\hbar}{2m_0\omega_0}\right)^3 \left[-(3n^2 + 3n + 2) - 9(3n^2 + 3n + 1)\right]$$

$$= \frac{\gamma^2}{\hbar\omega_0} \left(\frac{\hbar}{2m_0\omega_0}\right)^3 \left(-30n^2 - 30n - 11\right)$$

The energy shift is negative and grows as n^2. No matter how small γ, the energy shift eventually becomes larger than the level spacing, and perturbation theory breaks down.

This result can be understood by looking at the potential, where we assume both κ and γ are positive,[15]

$$V(x) = \frac{1}{2}\kappa x^2 + \gamma x^3$$

This is a cubic in x with the following properties:

- As $x \to \infty$, the potential goes to $+\infty$;
- As $x \to -\infty$, the potential goes to $-\infty$. Hence *there is no ground state*, and all the levels are actually unstable;
- The local extrema of the potential are identified from

$$\frac{dV(x)}{dx} = \kappa x + 3\gamma x^2 = 0$$

- There is a local minimum at $x = 0$ where $V(0) = 0$;
- There is a local maximum at $x = -\kappa/3\gamma$ with $V(-\kappa/3\gamma) = \kappa^3/54\gamma^2$. Below this maximum, the levels are bound, but unstable (by barrier penetration). Above this energy, the levels are actually unbound!

Problem 4.17 Show the case where $E = E_n^0$ in Eq. (4.47) can be handled by adding another perturbation H'', diagonal in degenerate subspaces, and in the end taking the limit $H'' \to 0$.

Solution to Problem 4.17

The expansion coefficients in the wave function in Eq. (4.47) are

$$a_n = \frac{\langle \phi_n | H' | \psi \rangle}{E - E_n^0} \qquad ; E \neq E_n^0$$

---

[15]Recall $\kappa = m_0\omega_0^2$.

Suppose the energy of the level is not moved by the interaction H', and $E = E_n^0$. Add a tiny additional perturbation $\lambda H''$ that *does* move the level, so that now $E \neq E_n^0$. Choose this additional perturbation so that it is diagonal in any subspace degenerate with ϕ_n. Equations (4.56) then read

$$E - E_n^0 = \langle \phi_n | H' + \lambda H'' | \psi \rangle$$

$$\psi(x) = \phi_n(x) + \sum_{m \neq n} \phi_m(x) \frac{\langle \phi_m | H' + \lambda H'' | \psi \rangle}{E - E_m^0}$$

We now observe that

- The limit $\lambda \to 0$ reproduces the results for non-degenerate perturbation theory in Eqs. (4.58)

$$\psi_n^{(1)}(x) = \phi_n(x) + \sum_{m \neq n} \phi_m(x) \frac{\langle \phi_m | H' | \phi_n \rangle}{E_n^0 - E_m^0}$$

$$E_n^{(1)} - E_n^0 = \langle \phi_n | H' | \phi_n \rangle$$

- Consider degenerate perturbation theory-I, where it is assumed that H' is also diagonal in the degenerate subspace so that

$$\langle \phi_m | H' + \lambda H'' | \phi_n \rangle = \delta_{mn} \langle \phi_n | H' + \lambda H'' | \phi_n \rangle$$
$$; \text{ all } (m, n) \text{ with } E_m^0 = E_n^0$$

The limit $\lambda \to 0$ then reproduces the results in Eqs. (4.67)

$$\psi_n^{(1)} = \phi_n + \sum_{m \neq n} \phi_m \frac{\langle \phi_m | H' | \phi_n \rangle}{E_n^0 - E_m^0} \qquad ; E_m^0 \neq E_n^0$$

$$E_n^{(1)} = E_n^0 + \langle \phi_n | H' | \phi_n \rangle$$

- The argument on diagonalizing $H' + \lambda H''$ in the degenerate subspace still holds, and therefore the results for degenerate perturbation-II in Eqs. (4.172) are also reproduced in the limit $\lambda \to 0$

$$\psi_s^{(1)} = \chi_s + \sum_{m \neq s} \frac{\phi_m \langle \phi_m | H' | \chi_s \rangle}{E_n^0 - E_m^0} \qquad ; E_m^0 \neq E_n^0$$

$$E_s^{(1)} = E_n^0 + \langle \chi_s | H' | \chi_s \rangle$$

Chapter 5

Scattering Theory

Problem 5.1 Show that the asymptotic form of the scattered wave ψ_{scatt} in Eq. (5.16) satisfies the time-independent Schrödinger equation as $r \to \infty$[1]

$$(\nabla^2 + k^2)\psi_{\text{scatt}} = 0 \qquad ; r \to \infty$$

Solution to Problem 5.1

The asymptotic form of the scattered wave in Eq. (5.16) is

$$\psi_{\text{scat}}(\mathbf{r}) = \frac{1}{\sqrt{V}} \left[f(\theta, \phi) \frac{e^{ikr}}{r} \right] \qquad ; r \to \infty$$

The laplacian in spherical coordinates can be written as[2]

$$\nabla^2 = \frac{1}{r} \frac{\partial^2}{\partial r^2} r + \frac{1}{r^2} \left[\frac{1}{\sin\theta} \frac{\partial}{\partial\theta} \sin\theta \frac{\partial}{\partial\theta} + \frac{1}{\sin^2\theta} \frac{\partial^2}{\partial\phi^2} \right]$$

When applied to ψ_{scat}, the last term makes a negligible contribution of $O(1/r^3)$. The first term gives

$$\frac{1}{r} \frac{\partial^2}{\partial r^2} r \frac{e^{ikr}}{r} = -k^2 \frac{e^{ikr}}{r}$$

Hence

$$(\nabla^2 + k^2)\psi_{\text{scatt}} = (-k^2 + k^2) \frac{1}{\sqrt{V}} \left[f(\theta, \phi) \frac{e^{ikr}}{r} \right] \qquad ; r \to \infty$$

$$= 0$$

---

[1] *Hint:* Recall Eq. (2.200).
[2] See Eqs. (2.190)–(2.191).

Problem 5.2 This problem provides an alternate derivation of the scattering Green's function. Define $\mathbf{r} \equiv \mathbf{x} - \mathbf{y}$, and introduce a new coordinate system whose origin is located at $\mathbf{y}$. Now consider the function $G_k^{(+)} \equiv e^{ikr}/4\pi r$.

(a) Show that for $r > 0$, this function satisfies

$$(\nabla^2 + k^2)\frac{e^{ikr}}{4\pi r} = 0 \qquad ; r > 0$$

(b) Take a small sphere of radius R about the origin, and use Gauss's law to show

$$\int_V d^3r\, \nabla^2 \frac{e^{ikr}}{4\pi r} = \int_A d\mathbf{A} \cdot \nabla \frac{e^{ikr}}{4\pi r}$$
$$= -1 \qquad ; R \to 0$$

where $d\mathbf{A} = \mathbf{e}_r R^2 d\Omega$ is a little element of surface area, and the last equality holds in the limit $R \to 0$. Hence conclude that the differential Eq. (5.19) defining the Green's function with a Dirac delta-function source is satisfied;

(c) Show that the same derivation holds for $G_k^{(-)} \equiv e^{-ikr}/4\pi r$, and thus conclude that the proper choice of Green's function depends on the imposed boundary conditions.

Solution to Problem 5.2

(a) The calculation here is exactly that of Prob. 5.1.

$$(\nabla^2 + k^2)\frac{e^{ikr}}{4\pi r} = \left(\frac{1}{r}\frac{\partial^2}{\partial r^2} r + k^2\right)\frac{e^{ikr}}{4\pi r} \qquad ; r > 0$$
$$= (-k^2 + k^2)\frac{e^{ikr}}{4\pi r} = 0$$

(b) We now show there is something left over when the first term in the above is integrated over a small sphere surrounding the origin, no matter how small the sphere. Use Gauss's theorem to write

$$\int_V d^3r\, \nabla^2 \frac{e^{ikr}}{4\pi r} = \int_V d^3r\, \nabla \cdot \nabla \frac{e^{ikr}}{4\pi r}$$
$$= \int_A d\mathbf{A} \cdot \nabla \frac{e^{ikr}}{4\pi r}$$

Here $d\mathbf{A} = \mathbf{e}_r R^2 d\Omega$ is a little element of surface area. The radial component of the gradient is $\nabla \doteq \mathbf{e}_r\, \partial/\partial r$. Differentiation of the exponent provides a vanishingly small contribution as $R \to 0$, and hence

$$\int_V d^3r \, \nabla^2 \frac{e^{ikr}}{4\pi r} = \int_A d\mathbf{A} \cdot \nabla \frac{e^{ikr}}{4\pi r}$$

$$= -\int_A d\mathbf{A} \cdot \mathbf{e}_r \left(\frac{e^{ikR}}{4\pi R^2} \right)$$

$$= -1 \qquad ; R \to 0$$

The second term on the l.h.s. in part (a) also makes a vanishingly small contribution when integrated over the sphere in the limit $R \to 0$. Therefore we can write

$$(\nabla^2 + k^2) \frac{e^{ikr}}{4\pi r} = -\delta^{(3)}(\mathbf{r})$$

This is the differential equation for the Green's function.

(c) Since exactly the same analysis holds for $e^{-ikr}/4\pi r$, we conclude that the proper choice of Green's function depends on the imposed boundary conditions.

Problem 5.3 (a) Carry out a calculation of $G_k^{(-)}(\mathbf{x}, \mathbf{y})$ in parallel with that for $G_k^{(+)}(\mathbf{x}, \mathbf{y})$ as detailed in the text, evaluate the residue of the pole at $t = -(k - i\eta)$, and hence derive Eq. (5.32);

(b) Show the contribution from the large semi-circle in Fig. 5.3 in the text makes a vanishing contribution to the contour integral in Eq. (5.29) as $R \to \infty$.[3]

Solution to Problem 5.3

(a) The scattering Green's function $G_k^{(-)}(\mathbf{x}, \mathbf{y})$ is defined in Eq. (5.29)

$$G_k^{(-)}(\mathbf{x}, \mathbf{y}) = \frac{1}{4\pi^2 i |\mathbf{x} - \mathbf{y}|} \oint_C \frac{t \, dt}{t^2 - (k - i\eta)^2} e^{it|\mathbf{x} - \mathbf{y}|} \qquad ; R \to \infty$$

where the contour C is shown in Fig. 5.3 in the text. Now the position of the poles is reflected with respect to the $y = 0$ axis, and it is the pole on the *left* that lies within the contour. A repetition of the argument in

---

[3]See [Fetter and Walecka (2003a)].

Eqs. (5.30) then gives

$$G_k^{(-)}(\mathbf{x}, \mathbf{y}) = \frac{1}{4\pi^2 i |\mathbf{x} - \mathbf{y}|} 2\pi i \operatorname{Res}\left[\frac{t}{t^2 - (k - i\eta)^2} e^{it|\mathbf{x} - \mathbf{y}|}\right]_{t=-k+i\eta}$$

$$= \frac{1}{4\pi^2 i |\mathbf{x} - \mathbf{y}|} 2\pi i \left[\frac{-k}{-2k} e^{-ik|\mathbf{x} - \mathbf{y}|}\right]$$

Therefore

$$G_k^{(-)}(\mathbf{x}, \mathbf{y}) = \frac{e^{-ik|\mathbf{x} - \mathbf{y}|}}{4\pi |\mathbf{x} - \mathbf{y}|}$$

(b) Consider the contribution to the contour integral from the semi-circle when R is very large. For t on that semi-circle, write

$$t = Re^{i\phi}$$
$$dt = iRe^{i\phi} d\phi$$

The contribution from the semi-circle then takes the form

$$\int_R \frac{t \, dt}{t^2 - (k - i\eta)^2} e^{it|\mathbf{x} - \mathbf{y}|} = i \int_0^\pi d\phi \, \frac{R^2 e^{2i\phi}}{R^2 e^{2i\phi} - (k - i\eta)^2} e^{iR(\cos\phi + i\sin\phi)|\mathbf{x} - \mathbf{y}|}$$

If $R \gg k$, then the $(k - i\eta)^2$ in the denominator can be neglected, and

$$\int_R \frac{t \, dt}{t^2 - (k - i\eta)^2} e^{it|\mathbf{x} - \mathbf{y}|} \approx i \int_0^\pi d\phi \, e^{iR(\cos\phi + i\sin\phi)|\mathbf{x} - \mathbf{y}|}$$

This expression is bounded by

$$\left|\int_R \frac{t \, dt}{t^2 - (k - i\eta)^2} e^{it|\mathbf{x} - \mathbf{y}|}\right| \leq \int_0^\pi d\phi \, e^{-R\sin\phi |\mathbf{x} - \mathbf{y}|}$$

The final integral is convergent and vanishes as $R \to \infty$ for any $|\mathbf{x} - \mathbf{y}| \neq 0$.

Problem 5.4 (a) Substitute the relation on the r.h.s. of Eq. (5.34) for $\psi_\mathbf{k}^{(+)}(\mathbf{y})$ in the integral, and derive the following *exact expression* for $\psi_\mathbf{k}^{(+)}(\mathbf{x})$

$$\psi_\mathbf{k}^{(+)}(\mathbf{x}) = e^{i\mathbf{k}\cdot\mathbf{x}} - \int d^3y \, G_k^{(+)}(\mathbf{x} - \mathbf{y})v(\mathbf{y})e^{i\mathbf{k}\cdot\mathbf{y}}$$

$$+ \int d^3y \int d^3z \, G_k^{(+)}(\mathbf{x} - \mathbf{y})v(\mathbf{y})G_k^{(+)}(\mathbf{y} - \mathbf{z})v(\mathbf{z})\psi_\mathbf{k}^{(+)}(\mathbf{z})$$

(b) Repeat this process once more to obtain an exact second-order iteration, with a final term explicitly of $O(v^3)$.

Solution to Problem 5.4

(a) Equations (5.34) provide an exact expression for $\Psi_{\mathbf{k}}^{(+)}(\mathbf{y})$ through

$$\psi_{\mathbf{k}}^{(+)}(\mathbf{y}) = e^{i\mathbf{k}\cdot\mathbf{y}} - \int d^3z \, G_k^{(+)}(\mathbf{y} - \mathbf{z})v(z)\psi_{\mathbf{k}}^{(+)}(\mathbf{z})$$

Substitution into Eq. (5.33) then gives

$$\psi_{\mathbf{k}}^{(+)}(\mathbf{x}) = e^{i\mathbf{k}\cdot\mathbf{x}} - \int d^3y \, G_k^{(+)}(\mathbf{x} - \mathbf{y})v(y)e^{i\mathbf{k}\cdot\mathbf{y}} +$$
$$\int d^3y \int d^3z \, G_k^{(+)}(\mathbf{x} - \mathbf{y})v(y)G_k^{(+)}(\mathbf{y} - \mathbf{z})v(z)\psi_{\mathbf{k}}^{(+)}(\mathbf{z})$$

(b) Substitution of the analogous expression for $\psi_{\mathbf{k}}^{(+)}(\mathbf{z})$ then gives

$$\psi_{\mathbf{k}}^{(+)}(\mathbf{x}) = e^{i\mathbf{k}\cdot\mathbf{x}} - \int d^3y \, G_k^{(+)}(\mathbf{x} - \mathbf{y})v(y)e^{i\mathbf{k}\cdot\mathbf{y}}$$
$$+ \int d^3y \int d^3z \, G_k^{(+)}(\mathbf{x} - \mathbf{y})v(y)G_k^{(+)}(\mathbf{y} - \mathbf{z})v(z)e^{i\mathbf{k}\cdot\mathbf{z}}$$
$$- \int d^3y \int d^3z \int d^3w \, G_k^{(+)}(\mathbf{x} - \mathbf{y})v(y)G_k^{(+)}(\mathbf{y} - \mathbf{z})v(z)$$
$$\times G_k^{(+)}(\mathbf{z} - \mathbf{w})v(w)\psi_{\mathbf{k}}^{(+)}(\mathbf{w})$$

This is still an exact integral equation for $\psi_{\mathbf{k}}^{(+)}(\mathbf{x})$, and the last term is explicitly of $O(v^3)$.

Problem 5.5 (a) What is the wavenumber k of a nucleon with energy $E_{\text{inc}} = 20\,\text{MeV}$ incident on a heavy object? With $100\,\text{MeV}$? With $500\,\text{MeV}$?

(b) What is the maximum momentum transfer q in elastic scattering in each case?

(c) If one were to measure the radius R of the potential through the location of the first minimum in Fig. 5.9 in the text, what is the smallest value of R that could be measured in each case?

Solution to Problem 5.5

(a) Use

$$\frac{\hbar^2}{2m_p} = 20.74\,\text{MeV-F}^2 \qquad ; 1\,\text{F} = 10^{-13}\,\text{cm}$$

Then, with $E = \hbar^2 k^2 / 2m_p$, one has

$$k = \left(\frac{2m_p E}{\hbar^2} \right)^{1/2}$$

The values of k in F^{-1} for energies of $E = 20\,\mathrm{MeV}, 100\,\mathrm{MeV}$, and $500\,\mathrm{MeV}$ are shown in the second column of Table 5.1.

(b) The maximum momentum transfer is achieved through back scattering where in elastic scattering $q_{max} = 2k$. The corresponding values of q_{max} in F^{-1} are shown in the third column in Table 5.1.

(c) The first zero in Fig. 5.9 in the text occurs at

$$qR = 4.493$$

If the location of this zero is used to determine R, the minimum value $R_{min} = 4.493/q_{max}$ that can be observed in each case is given in F in the fourth column in Table 5.1.

Table 5.1   Incident energy, incident wave number, maximum momentum transfer, and corresponding minimum radius observable from the diffraction minimum in Prob. 5.5.

$E(\mathrm{MeV})$	$k(\mathrm{F}^{-1})$	$q_{max}(\mathrm{F}^{-1})$	$R_{min}(\mathrm{F})$
20	0.982	1.964	2.288
100	2.196	4.392	1.023
500	4.910	9.820	0.457

Problem 5.6 Calculate the scattering amplitude and differential cross section in Born approximation for the exponential potential, with either sign of V_0,

$$V(r) = V_0\, e^{-\lambda r} \qquad ;\ \text{exponential potential}$$

Solution to Problem 5.6 -

The required Fourier transform of the potential can be simply obtained as the negative of the derivative with respect to λ of the result for the

Yukawa potential in Eqs. (5.59)–(5.62)[4]

$$\tilde{V}(q) = -\frac{d}{d\lambda}\int d^3x\, e^{-i\mathbf{q}\cdot\mathbf{x}}\frac{V_0 e^{-\lambda x}}{x} = -\frac{d}{d\lambda}\frac{4\pi V_0}{q^2 + \lambda^2}$$

$$= \frac{8\pi\lambda V_0}{(q^2 + \lambda^2)^2}$$

This gives a Born-approximation scattering amplitude and differential cross section of

$$f_{\mathrm{BA}}(q) = -\frac{2\mu}{\hbar^2}\frac{2\lambda V_0}{(q^2 + \lambda^2)^2}$$

$$\left(\frac{d\sigma}{d\Omega}\right)_{\mathrm{BA}} = |f_{\mathrm{BA}}(q)|^2$$

Problem 5.7 (a) Given a real potential $v(|\mathbf{x}|)$, and the asymptotic form of the wave function in the first of Eqs. (5.38), use the continuity equation and Gauss's law to derive the optical theorem

$$\frac{4\pi}{k}\mathrm{Im}\,f_{\mathrm{el}}(0) = \sigma_{\mathrm{elastic}} \qquad ; \text{ optical theorem}$$

where $\sigma_{\mathrm{elastic}}$ is the integrated elastic cross section;
(b) Generalize this result to complex v [see Eq. (5.247)]. Show[5]

$$\frac{4\pi}{k}\mathrm{Im}\,f_{\mathrm{el}}(k, 0) = \sigma_{\mathrm{total}} \qquad ; \text{ optical theorem}$$

where σ_{total} is the *total* cross section, now including an absorptive part.

Solution to Problem 5.7

(a) We start with the continuity Eq. (2.6)

$$\frac{\partial\rho}{\partial t} + \boldsymbol{\nabla}\cdot\mathbf{S} = 0$$

where ρ is the probability density, and $\mathbf{S}$ is the probability current

$$\rho = |\psi(\mathbf{x}, t)|^2$$

$$\mathbf{S} = \frac{\hbar}{2im}\left[\psi^\star\,\boldsymbol{\nabla}\psi - (\boldsymbol{\nabla}\psi)^\star\,\psi\right]$$

---

[4]Here $x = |\mathbf{x}| = r$.
[5]Note Eq. (20.3) in [Schiff (1968)].

If we integrate the continuity equation over space, and use Gauss's law, we have

$$-\frac{d}{dt}\int_V \rho \, d^3x = \int_A \mathbf{S}\cdot d\mathbf{A}$$

where A is the surface enclosing the volume V. Clearly, if one chooses V to be a sphere of radius R and lets $R \to \infty$, then the l.h.s. of the equation above vanishes (since the total probability is constant in this case). Hence

$$\int_A \mathbf{S}\cdot d\mathbf{A} = 0$$

where $\mathbf{S}$ is now evaluated at the surface of the infinite sphere, with $d\mathbf{A} = \mathbf{e}_r R^2 \sin\theta \, d\theta \, d\phi$.

We now need to use Eq. (5.38) for the total scattering wave function[6]

$$\psi_{\mathbf{k}}^{(+)}(\mathbf{x}) \to e^{i\mathbf{k}\cdot\mathbf{x}} + f(\theta,\phi)\frac{e^{ikx}}{x}$$
$$\equiv \psi_{\text{inc}}(\mathbf{x}) + \psi_{\text{scat}}(\mathbf{x})$$

where it has been decomposed into an incident and scattered parts. Then

$$\mathbf{S} = \frac{\hbar}{2im}\left[\psi^\star\boldsymbol{\nabla}\psi - (\boldsymbol{\nabla}\psi)^\star\psi\right]$$
$$= \frac{\hbar}{2im}\left[\psi_{\text{inc}}^\star\boldsymbol{\nabla}\psi_{\text{inc}} - (\boldsymbol{\nabla}\psi_{\text{inc}})^\star\psi_{\text{inc}}\right] + \frac{\hbar}{2im}\left[\psi_{\text{scat}}^\star\boldsymbol{\nabla}\psi_{\text{scat}} - (\boldsymbol{\nabla}\psi_{\text{scat}})^\star\psi_{\text{scat}}\right]$$
$$+\frac{\hbar}{2im}\left[\psi_{\text{inc}}^\star\boldsymbol{\nabla}\psi_{\text{scat}} + \psi_{\text{scat}}^\star\boldsymbol{\nabla}\psi_{\text{inc}} - (\boldsymbol{\nabla}\psi_{\text{scat}})^\star\psi_{\text{inc}} - (\boldsymbol{\nabla}\psi_{\text{inc}})^\star\psi_{\text{scat}}\right]$$

We start considering

$$\mathbf{S}_{\text{inc}}\cdot d\mathbf{A} = \frac{\hbar}{2im}\left[\psi_{\text{inc}}^\star\boldsymbol{\nabla}\psi_{\text{inc}} - (\boldsymbol{\nabla}\psi_{\text{inc}})^\star\psi_{\text{inc}}\right]\cdot d\mathbf{A}$$
$$= \frac{\hbar k\cos\theta}{m}R^2 \sin\theta \, d\theta \, d\phi$$

In this case the surface integral vanishes

$$\int \mathbf{S}_{\text{inc}}\cdot d\mathbf{A} = 0$$

With a plane wave, there is cancelling incoming and outgoing flux over the sphere.

---

[6]Recall that here $x = |\mathbf{x}| = r$.

Next, we consider

$$\mathbf{S}_{\text{scat}} \cdot d\mathbf{A} = \frac{\hbar}{2im} \left[\psi_{\text{scat}}^\star \boldsymbol{\nabla} \psi_{\text{scat}} - (\boldsymbol{\nabla} \psi_{\text{scat}})^\star \psi_{\text{scat}} \right] \cdot d\mathbf{A}$$

$$= \frac{\hbar k}{m} \frac{|f|^2}{R^2} R^2 \sin\theta \, d\theta \, d\phi$$

The factors of R^2 cancel, and in this case the surface integral becomes

$$\int \mathbf{S}_{\text{scat}} \cdot d\mathbf{A} = \frac{\hbar k}{m} \int |f|^2 \, d\Omega = \frac{\hbar k}{m} \sigma$$

Finally, we need to calculate the interference contribution

$$\mathbf{S}_{\text{int}} \cdot d\mathbf{A} = \frac{\hbar}{2im} \left[\psi_{\text{inc}}^\star \frac{\partial \psi_{\text{scat}}}{\partial r} + \psi_{\text{scat}}^\star \frac{\partial \psi_{\text{inc}}}{\partial r} - \left(\frac{\partial \psi_{\text{scat}}}{\partial r} \right)^\star \psi_{\text{inc}} \right.$$

$$\left. - \left(\frac{\partial \psi_{\text{inc}}}{\partial r} \right)^\star \psi_{\text{scat}} \right]_R dA$$

$$= \frac{\hbar}{m} \text{Im} \left[\psi_{\text{inc}}^\star \frac{\partial \psi_{\text{scat}}}{\partial r} + \psi_{\text{scat}}^\star \frac{\partial \psi_{\text{inc}}}{\partial r} \right]_R dA$$

$$= \frac{\hbar}{mR^2} \text{Im} \left[(ikR - 1) \, f(\theta, \phi) \, e^{ikR(1 - \cos\theta)} \right.$$

$$\left. + ikR \cos\theta \, f^\star(\theta, \phi) \, e^{-ikR(1 - \cos\theta)} \right] R^2 d\cos\theta \, d\phi$$

Note that the factors of R^2 again cancel.

Let us first discuss the integral over ϕ: because of the periodic boundary conditions one can express $f(\theta, \phi)$ as a Fourier series in ϕ, and it is straightforward to see that the integral over ϕ picks the constant term in this series, that is, the term independent of ϕ. For this reason we will eliminate the ϕ dependence in the expressions of $f(\theta)$.[7] We then use

$$\int_0^{2\pi} d\phi = 2\pi$$

Now we concentrate on the integral over $d\cos\theta$. Note that the exponentials in the square brackets are functions of

$$u \equiv kR(1 - \cos\theta)$$

In the integral, as $kR \to \infty$, the phase of the exponential oscillates rapidly unless $(1 - \cos\theta) \approx 0$, and the contribution to the integral then all comes from the forward direction, where the phase is stationary. Without loss

[7]Compare Eq. (5.106).

of generality, we may consider f to be a function of $1 - \cos\theta$ $[f(\theta) = \tilde{f}(1 - \cos\theta)]$, and the surface integral may be cast into the general form

$$\int_{-1}^{1} d\cos\theta\, g[kR(1 - \cos\theta)]\tilde{f}(1 - \cos\theta) = \int_{0}^{2kR} du\, \frac{g(u)}{kR}\, \tilde{f}\left(\frac{u}{kR}\right)$$

where the explicit form of g will depend on whether we are considering the real or imaginary part of $f(\theta)$. Let us then separate the contributions from those real and imaginary parts

$$\begin{aligned}
\mathbf{S}_{\text{int}} \cdot d\mathbf{A} = {} & \frac{\hbar}{m}\operatorname{Re} f(\theta)\{kR\cos\left[kR(1 - \cos\theta)\right] - \sin\left[kR(1 - \cos\theta)\right] \\
& + kR\cos\theta\cos\left[kR(1 - \cos\theta)\right]\}\, d\cos\theta d\phi \\
& + \frac{\hbar}{m}\operatorname{Im} f(\theta)\{-\cos\left[kR(1 - \cos\theta)\right] - kR\sin\left[kR(1 - \cos\theta)\right] \\
& - kR\cos\theta\sin\left[kR(1 - \cos\theta)\right]\}\, d\cos\theta d\phi
\end{aligned}$$

With the expectation that as $R \to \infty$, the dominant contribution to the integral comes from the forward direction where $u = 0$, we proceed to approximate $\tilde{f}(u/kR)$ with its Taylor expansion around the origin.

Let us examine separately the two cases: in the first case, where we consider the expression containing $\operatorname{Re} f(\theta)$ we have

$$g(u) \equiv \frac{\hbar}{m}[(2kR - u)\cos u - \sin u]$$

and

$$\frac{1}{kR}\int_{0}^{2kR} g(u)\, du = 0$$

$$\frac{1}{kR}\int_{0}^{2kR} g(u)\left(\frac{u}{kR}\right) du = O\left(\frac{1}{R}\right)$$

$$\frac{1}{kR}\int_{0}^{2kR} g(u)\left(\frac{u}{kR}\right)^{2} du = O\left(\frac{1}{R^{2}}\right) \qquad ; \text{ etc.}$$

As a result, the contribution containing $\operatorname{Re} f(\theta)$ vanishes when $R \to \infty$.

Next, we consider the contribution containing $\operatorname{Im} f(\theta)$: in this case

$$g(u) \equiv \frac{\hbar}{m}[(u - 2kR)\sin u - \cos u]$$

and

$$\frac{1}{kR} \int_0^{2kR} g(u)\, du = -\frac{2\hbar}{m}$$

$$\frac{1}{kR} \int_0^{2kR} g(u) \left(\frac{u}{kR}\right) du = O\left(\frac{1}{R^2}\right)$$

$$\frac{1}{kR} \int_0^{2kR} g(u) \left(\frac{u}{kR}\right)^2 du = O\left(\frac{1}{R^2}\right) \qquad ; \text{ etc.}$$

Once the limit $R \to \infty$ is taken, one can see that only the first term corresponding to $u = 0$ (or, equivalently, $\theta = 0$) survives. Therefore

$$\lim_{R\to\infty} \int \mathbf{S}_{\text{int}} \cdot d\mathbf{A} = -\frac{4\pi\hbar}{m} \operatorname{Im} f(0)$$

where a factor 2π comes from the integration over ϕ.

With the use of these results inside the continuity equation, we obtain the optical theorem

$$\frac{4\pi}{k} \operatorname{Im} f_{\text{el}}(0) = \sigma_{\text{elastic}} \qquad ; \text{ optical theorem}$$

(b) In the presence of a complex potential, the continuity equation is modified to

$$\frac{\partial \rho}{\partial t} + \nabla \cdot \mathbf{S} = -\frac{2\rho}{\hbar} \operatorname{Im} V$$

where $V(|\mathbf{x}|) = \operatorname{Re} V - i \operatorname{Im} V$. This equation may be integrated over space, as done before, obtaining

$$-\frac{d}{dt} \int_V \rho\, d^3x = \int_A \mathbf{S} \cdot d\mathbf{A} + \frac{2}{\hbar} \int d^3x\, \rho \operatorname{Im} V$$

The absorption cross section is defined to be the absorption probability rate divided by the incident probability flux, which in this case is $I_{\text{inc}} = \hbar k / m$. Thus[8]

$$\sigma_{\text{abs}} = \frac{2m}{\hbar^2 k} \int d^3x\, \rho \operatorname{Im} V$$

Hence we obtain [compare Prob. 5.20(a)]

$$\frac{4\pi}{k} \operatorname{Im} f_{\text{el}}(k, 0) = \sigma_{\text{total}} \qquad ; \text{ optical theorem}$$

where $\sigma_{\text{total}} = \sigma_{\text{el}} + \sigma_{\text{abs}}$.

---

[8]See Eq. (20.3) of [Schiff (1968)].

Problem 5.8 (a) Show that $G_k^{(\pm)}(\mathbf{x}-\mathbf{y})$ reproduce the Coulomb Green's function in the limit $k \to 0$;

(b) Take the limit $k \to 0$ in Eq. (5.79), and derive the generating function for the Legendre polynomials in Eq. (2.66).

Solution to Problem 5.8

(a) The first part follows directly from

$$\text{Lim}_{k\to 0} \frac{e^{ik|\mathbf{x}-\mathbf{y}|}}{4\pi|\mathbf{x}-\mathbf{y}|} = \frac{1}{4\pi|\mathbf{x}-\mathbf{y}|}$$

(b) The second part follows through the use of Eqs. (5.124) in taking the $k \to 0$ limit of the expressions in Eqs. (5.84) and (5.101)

$$\frac{1}{4\pi|\mathbf{x}-\mathbf{y}|} = \frac{1}{4\pi}\sum_{l=0}^{\infty}\frac{1}{x_>}\left(\frac{x_<}{x_>}\right)^l P_l(\cos\theta_{xy})$$

This is just the the generating function for the Legendre polynomials in Eq. (2.66).

Problem 5.9 Start from the expression for the scattering amplitude in Eq. (5.38). Insert the plane-wave expansion in Eq. (5.78), the partial-wave expansion of the scattering state in Eq. (5.96), and use the addition theorem to re-derive the partial-wave decomposition of the scattering amplitude in Eq. (5.105).[9]

Solution to Problem 5.9

The starting point is the expression for the scattering amplitude and the expansion of the scattering state in Eqs. (5.38) and (5.96)

$$f(\theta, \phi) \equiv -\frac{1}{4\pi}\int d^3y\, e^{-i\mathbf{k}'\cdot\mathbf{y}}\, v(y)\psi_{\mathbf{k}}^{(+)}(\mathbf{y})$$

$$\psi_{\mathbf{k}}^{(+)}(\mathbf{x}) = \sum_{l=0}^{\infty}(2l+1)i^l\psi_l^{(+)}(x;k)P_l(\cos\theta_{kx})$$

These are to be analyzed with the aid of the plane-wave expansion and the

---

[9] Recall $\mathbf{x} = x\mathbf{e}_x$ and $\mathbf{k}' = k\mathbf{e}_x$.

addition theorem in Eqs. (5.78) and (5.77)

$$e^{i\mathbf{k}\cdot\mathbf{x}} = \sum_{l=0}^{\infty} \sum_{m=-l}^{l} 4\pi i^l j_l(kx) Y_{lm}(\Omega_k) Y_{lm}^{\star}(\Omega_x)$$

$$P_l(\cos\theta_{12}) = \frac{4\pi}{2l+1} \sum_{m=-l}^{l} Y_{lm}(\Omega_1) Y_{lm}^{\star}(\Omega_2)$$

Insert these last two expressions in $f(\theta, \phi)$, noting that $\mathbf{k}' = k\mathbf{e}_x$,

$$f(\theta, \phi) = -\frac{1}{4\pi} \int y^2 dy d\Omega_y \sum_l \sum_m 4\pi(-i)^l j_l(ky) Y_{lm}(\Omega_x) Y_{lm}^{\star}(\Omega_y)$$

$$\times v(y) \sum_{l'} \sum_{m'} 4\pi i^{l'} \psi_{l'}^{(+)}(y; k) Y_{l'm'}(\Omega_y) Y_{l'm'}^{\star}(\Omega_k)$$

Now use the orthonormality of the spherical harmonics

$$\int d\Omega_y \, Y_{lm}^{\star}(\Omega_y) Y_{l'm'}(\Omega_y) = \delta_{ll'} \delta_{mm'}$$

to obtain

$$f(\theta, \phi) = -\int y^2 dy \sum_l \sum_m 4\pi j_l(ky) v(y) \psi_l^{(+)}(y; k) Y_{lm}(\Omega_x) Y_{lm}^{\star}(\Omega_k)$$

Use of the addition theorem again then yields the expression in Eq. (5.105)

$$f(\theta, \phi) = -\sum_{l=0}^{\infty} (2l+1) P_l(\cos\theta_{kx}) \int_0^{\infty} y^2 \, dy \, j_l(ky) v(y) \psi_l^{(+)}(y; k)$$

Problem 5.10 (a) Locate a numerical program to calculate the differential cross section $d\sigma/d\Omega$ for scattering from a hard sphere, and reproduce the two curves shown in Fig. 5.1 for the cases $ka = 0.5$ and $ka = 10$;

(b) Extend the calculations in θ, and show that the integrated cross sections, in units of a^2, are 11.75 and 7.53 respectively;

(c) Extend the calculations as far as you can in ka.

Solution to Problem 5.10

(a) The phase shift for scattering from a hard sphere has been obtained in Eq. (5.140) of the text

$$\tan\delta_l = \frac{j_l(ka)}{n_l(ka)}$$

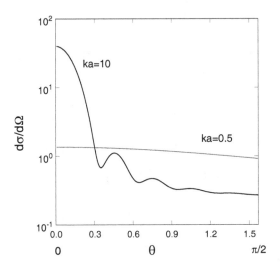

Fig. 5.1   Cross section $d\sigma/d\Omega$, measured in units of a^2, for scattering from a hard sphere of radius a. Two cases are plotted: $ka = 0.5$ and $ka = 10$. Note that the plot only displays the region $0 \le \theta \le \pi/2$.

Notice that this result is exact and holds for all l and k.

The scattering amplitude is expressed in terms of the phase shift in Eq. (5.119) of the text

$$f(k, \theta) = \sum_{l=0}^{\infty} (2l + 1) \frac{e^{i\delta_l(k)} \sin \delta_l(k)}{k} P_l(\cos \theta)$$

The differential cross section is given by the absolute square of the scattering amplitude [see Eq. (5.18)]

$$\frac{d\sigma}{d\Omega} = |f(k, \theta)|^2$$

In Fig. 5.2 we reproduce the plot of Fig. 5.1 with a numerical calculation where the infinite series is approximated with the first 21 terms.

(b) The integrated cross section is

$$\sigma = \int \frac{d\sigma}{d\Omega} d\Omega = 2\pi \int_{-1}^{1} d\cos\theta \, |f(k, \theta)|^2$$

With the use of the orthogonality relation for the Legendre polynomials,

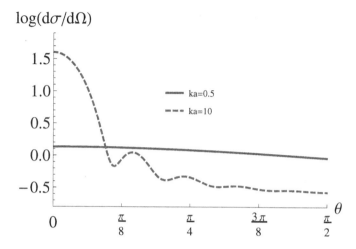

Fig. 5.2 Numerical reproduction of Fig. 5.1 retaining the first 21 terms in the partial wave expansion.

one obtains

$$\sigma = \frac{4\pi}{k^2} \sum_{l=0}^{\infty} (2l+1) \sin^2 \delta_l = \frac{4\pi}{k^2} \sum_{l=0}^{\infty} (2l+1) \frac{j_l^2(ka)}{j_l^2(ka) + n_l^2(ka)}$$

Although one can obtain the value of the integrated cross section at $ka = 0.5$ and $ka = 10$ by simply substituting the numerical values in the above expression and summing over a sufficiently large number of terms, it is instructive to look at the behavior of the summands

$$\sigma_l \equiv \frac{4\pi}{k^2}(2l+1) \frac{j_l^2(ka)}{j_l^2(ka) + n_l^2(ka)}$$

which represent the contribution of the lth wave to the integrated cross section.

It is useful at this point to remember the explicit expressions for the spherical Bessel functions appearing in our calculations

$$j_l(x) = (-1)^l x^l \left(\frac{1}{x}\frac{d}{dx} \right)^l \frac{\sin x}{x}$$

$$n_l(x) = (-1)^{l+1} x^l \left(\frac{1}{x}\frac{d}{dx} \right)^l \frac{\cos x}{x}$$

These relations can be used to obtain the asymptotic behavior of the spherical Bessel functions for $x \ll l$ and $x \gg l$:

For $x \ll l$ (that is, the argument much less than the order), one has

$$j_l(x) \approx \frac{x^l}{(2l+1)!!}$$

$$n_l(x) \approx -\frac{(2l-1)!!}{x^{l+1}}$$

For $x \gg l$ (that is, the argument much greater than the order), one has[10]

$$j_l(x) \approx \frac{1}{x} \sin\left(x - l\frac{\pi}{2}\right)$$

$$n_l(x) \approx -\frac{1}{x} \cos\left(x - l\frac{\pi}{2}\right)$$

With the use of these relations, we see that for $ka \ll l$ the term corresponding to $l = 0$ dominates in the series, whereas for $ka \gg l$ the summands corresponding to different values of l are comparable in size: therefore, for a given value of ka, one can obtain a good approximation with the partial sum containing the terms up to $l_{max} \approx \lceil ka \rceil$ (here $\lceil x \rceil$ is the smallest integer which is larger than, or equal to x).

With this approximation, we obtain for $ka = 0.5$

$$\sigma/a^2 \approx 11.75$$

which agrees with the value quoted in this problem.

For $ka = 10$ (corresponding to $l_{max} = 10$), one obtains

$$\sigma/a^2 \approx 7.47$$

which differs by less than 1% from the value quoted in this problem. With two more extra terms in the sum, corresponding to $l_{max} = 12$, one has

$$\sigma/a^2 \approx 7.53$$

which now agrees with the result quoted in the text.

In Fig. 5.3 we display the summands σ_l/a^2 for $ka = 0.5$ and $ka = 10$ for different values of l: this plot should convince the reader of the possibility of using a cutoff $l_{max} \approx ka$ in the series.

---

[10]The asymptotic relations given in the text are [Eqs. (5.135)–(5.136)]

$$j_l(x) \rightarrow \frac{1}{x} \cos\left[x - (l+1)\pi/2\right] \qquad ; x \rightarrow \infty$$

$$n_l(x) \rightarrow \frac{1}{x} \sin\left[x - (l+1)\pi/2\right]$$

See [Fetter and Walecka (2003a)].

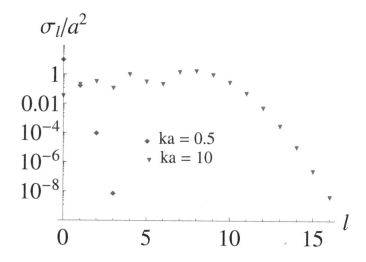

Fig. 5.3 σ_l/a^2 for $ka = 0.5$ and $ka = 10$ as a function of l.

(c) As we have discussed in part (b), the evaluation of the integrated cross section at a given value of ka requires one to sum the first $l_{\max} + 1$ terms, with $l_{\max} \approx \lceil ka \rceil + 2$. Therefore as ka grows, a larger number of terms needs to be taken into account, and the numerical calculation gets slower. For example, for $ka = 10^3$, we need to evaluate the first 1003 terms in the series and then sum them; the result is

$$\sigma/a^2 \approx 6.34$$

while for $ka = 5 \times 10^3$, the sum provides

$$\sigma/a^2 \approx 6.30$$

Notice, however, that the time needed by the computer in evaluating the sums grows much faster than linear (in our case it took about 60 times more time to evaluate σ for $ka = 5000$ than for $ka = 1000$), and therefore one should be careful in choosing large, but not too large, values of ka for the numerical calculations.

In contrast, we may quite simply calculate the asymptotic value of σ for $ka \to \infty$ analytically. We just observe that the spherical Bessel functions obey the recurrence relations

$$j_l(x) = \frac{2l-1}{x} \, j_{l-1}(x) - j_{l-2}(x)$$

$$n_l(x) = \frac{2l-1}{x} \, n_{l-1}(x) - n_{l-2}(x)$$

For $x \to \infty$ these relations may be approximated to

$$j_l(x) \approx -j_{l-2}(x) + \cdots$$

$$n_l(x) \approx -n_{l-2}(x) + \cdots$$

Clearly, if we repeatedly apply these relations we obtain

$$j_l^2(x) \approx \begin{cases} x^{-2}\sin^2(x) & ; \, l \text{ even} \\ x^{-2}\cos^2(x) & ; \, l \text{ odd} \end{cases}$$

and

$$n_l^2(x) \approx \begin{cases} x^{-2}\cos^2(x) & ; \, l \text{ even} \\ x^{-2}\sin^2(x) & ; \, l \text{ odd} \end{cases}$$

Therefore, for $x \to \infty$

$$\frac{j_l^2(x)}{j_l^2(x) + n_l^2(x)} \approx \begin{cases} \sin^2(x) & ; \, l \text{ even} \\ \cos^2(x) & ; \, l \text{ odd} \end{cases}$$

With the use of this relation, we may write

$$\sigma \approx \frac{4\pi}{k^2} \left\{ \sum_{l=0}^{l_{\max}/2} (4l+1) \, \sin^2(ka) + \sum_{l=0}^{l_{\max}/2} (4l+3) \, \cos^2(ka) \right\}$$

$$\approx \frac{4\pi}{k^2} \left\{ 4 \int_0^{l_{\max}/2} l \, dl \right\}$$

This produces the asymptotic relation

$$\sigma \approx 2\pi a^2 \qquad\qquad ; \, ka \to \infty$$

where we have used $l_{\max} \approx ka$ in obtaining this result.

In Fig. 5.4 we have plotted the integrated cross section, in units of a^2, for $10^{-3} \le ka \le 10^3$ (solid line); the horizontal dashed line corresponds to $\lim_{k \to 0} \sigma/a^2 = 4\pi$ (see Prob. 5.11), while the horizontal dot-dashed line corresponds to $\lim_{k \to \infty} \sigma/a^2 = 2\pi$.

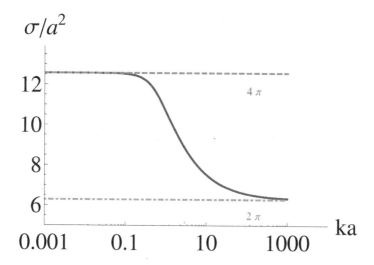

Fig. 5.4 Integrated cross section σ/a^2 as a function of ka for scattering from a hard sphere. The series is approximated using $l_{\max} \approx \lceil ka \rceil + 2$. The calculation of the asymptotic value 2π is discussed above.

Problem 5.11 (a) Show that at zero incident energy the differential cross section for scattering from a hard sphere of radius a is $d\sigma/d\Omega = a^2$, independent of angle;

(b) Show the total cross section is $\sigma = 4\pi a^2$;

(b) Compare with the numerical results in Prob. 5.10.

Solution to Problem 5.11

(a) It is clear from the result in Eq. (5.128) [and from the next problem] that the phase shifts for the hard sphere vanish in the zero-energy limit as

$$\delta_l(k) \sim (ka)^{2l+1} \qquad ; \, k \to 0$$

In this same limit, the scattering amplitude follows from Eqs. (5.119) and (5.144) as

$$f(k,\theta) \to \frac{\delta_0(k)}{k} P_0(\cos\theta) \qquad ; \, k \to 0$$
$$= -a$$

(b) The differential and total cross sections are then given by

$$\frac{d\sigma}{d\Omega} = a^2$$

$$\sigma = \int d\Omega \, \frac{d\sigma}{d\Omega} = 4\pi a^2$$

(c) In units of a^2, the total cross section at $ka = 0$ is $4\pi = 12.57$. This compares with the numerical results 11.75 and 7.53 found in Prob. 5.10(b) for the cases $ka = 0.5$ and $ka = 10$, respectively.

Problem 5.12 Show that as $k \to 0$, the phase shifts for scattering from a hard sphere of radius a behave as

$$\delta_l(k) \to -\frac{(2l+1)}{[(2l+1)!!]^2}(ka)^{2l+1} \qquad ; k \to 0$$

Solution to Problem 5.12

The phase shifts for scattering from a hard sphere are given by Eq. (5.140) as

$$\tan \delta_l(k) = \frac{j_l(ka)}{n_l(ka)}$$

As $ka \to 0$ at low energy, one can use the expressions in Eqs. (2.210)

$$j_l(z) \to \frac{z^l}{(2l+1)!!} \qquad ; n_l(z) \to -\frac{(2l-1)!!}{z^{l+1}} \qquad ; z \to 0$$

This gives[11]

$$\tan \delta_l(k) \to -\frac{(2l+1)}{[(2l+1)!!]^2}(ka)^{2l+1} \qquad ; ka \to 0$$

In this limit, the phase shift itself also goes to zero, and hence $\tan \delta_l(k) \to \delta_l(k)$. Thus, finally,

$$\delta_l(k) \to -\frac{(2l+1)}{[(2l+1)!!]^2}(ka)^{2l+1} \qquad ; ka \to 0$$

Problem 5.13 (a) Sketch the situation on the l.h.s. of Fig. 5.24 in the text when the attractive potential gives rise to a positive scattering length;

---

[11]Here we have written $1/(2l-1)!! = (2l+1)/(2l+1)!!$ to explicitly exhibit the correct $l = 0$ expression.

(b) When will the scattering length a become infinite?

(c) Explain how it is possible to have an infinite zero-energy cross section in quantum mechanics.

Solution to Problem 5.13

(a) It follows from Eq. (5.159) that outside the potential, the zero-energy scattering wave function is normalized to

$$\frac{u_0(r)}{k} \to r - a \qquad ; k \to 0 \qquad (5.1)$$

where a is the scattering length. Hence this solution rises with unit slope and extrapolates back to cross the axis at $r = a$. The more familiar situation is sketched in Fig. 5.24 in the text. A less familiar situation is the case where the potential is attractive, and the resulting scattering length is positive. Inside the potential the true wave function will be bent toward the axis, and the actual $u(r)$ must vanish at the origin. Hence, in this case, the actual solution matches on to the required asymptotic form as sketched in Fig. 5.5.

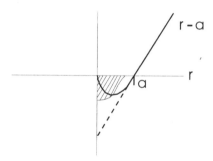

Fig. 5.5 Sketch of the situation with a positive scattering length a for an attractive potential with asymptotic zero-energy solution $r - a$. (Compare Fig. 5.24 in the text.)

(b) If the true solution matches on to the asymptotic solution with zero slope outside the potential, then the scattering length will be infinite (compare Fig. 2.13).

(c) The cross section is defined through the ratio of the probability of a given transition per unit time to the incident probability flux [see Eq. (6.92) and accompanying discussion]. While the former may be a finite quantity at zero energy, the latter goes to zero in this limit. Hence the cross section may be infinite in the zero-energy limit, as evidenced explicitly in the example in Eq. (5.285).

****

(*Aside*) We present one actual calculation of the situation for the attractive square-well potential in Fig. 5.23 in the text, with

$$v_0 = \frac{2\mu}{\hbar^2} V_0$$

For $k \to 0$ we have

$$\delta_0(k) = -ka$$

and, correspondingly, Eqs. (5.151) and (5.154) can be used to give

$$\frac{u_0(r)}{k} \approx \frac{1}{\sqrt{v_0}} \sin\left(r\sqrt{v_0}\right) \sec\left(R\sqrt{v_0}\right) \qquad ; r < R$$

$$= \frac{1}{k} \sin k(r-a) \approx r - a \qquad ; r \geq R$$

From Eq. (5.157) we have

$$\frac{a}{R} = 1 - \frac{1}{\sqrt{v_0}R\cot(\sqrt{v_0}R)}$$

This quantity is plotted in Fig. 5.6.   We see that the scattering length

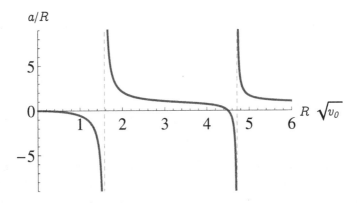

Fig. 5.6   Scattering length divided by R as a function of $\sqrt{v_0}R$.

is positive in a region $\pi/2 < \sqrt{v_0}R < 4.493$ (further regions of positivity appear at larger values of $\sqrt{v_0}R$). Moreover, for $\pi/2 < \sqrt{v_0}R < \pi$ we observe that $a > R$, whereas for $\pi < \sqrt{v_0}R < 4.493$ we have $a < R$.

In Figs. 5.7, 5.8, and 5.9 we plot $u_0(r)/k$ for $k \to 0$ in these three cases; the circled point marks the location $r = a$.

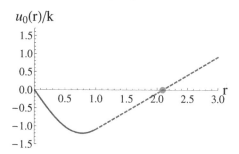

Fig. 5.7 $u_0(r)/k$ for $k \to 0$ with $R = 1$, which sets the length scale, and $v_0 = 4$.

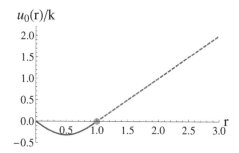

Fig. 5.8 $u_0(r)/k$ for $k \to 0$ with $R = 1$ and $v_0 = \pi^2$.

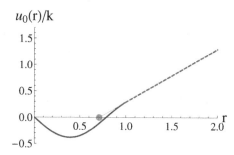

Fig. 5.9 $u_0(r)/k$ for $k \to 0$ with $R = 1$ and $v_0 = 9$.

The expression for the scattering length can be obtained using Eq. (5.164)

$$a = R - \frac{\tan\left(R\sqrt{v_0}\right)}{\sqrt{v_0}}$$

This becomes infinity for $\sqrt{v_0}R = \pi/2$. As we have seen in Prob. 2.20, for $\sqrt{v_0}R = \pi/2$ there is one bound state at zero energy. Thus, the divergence of the scattering length can be attributed to the presence of this bound state.

In Fig. 5.9 we plot $u_0(r)/k$ with $R = 1$ and $v_0 = (1.6)^2$, corresponding to $\sqrt{v_0}R = 1.6$, just above the critical value. In this case $a \approx 22.4$ (not shown in the plot). The wave function has a behavior similar to that shown in Fig. 2.13, except that it is here reflected through the $y = 0$ axis.

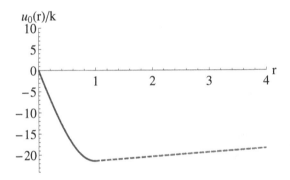

Fig. 5.10 $u_0(r)/k$ for $k \to 0$ with $R = 1$ and $v_0 = (1.6)^2$.

****

Problem 5.14 The condition that the attractive square-well potential in Fig. 5.23 in the text have just one bound state at zero energy was shown in Prob. 2.20 to be

$$\sqrt{v_0}R = \frac{\pi}{2} \qquad ; \text{ zero-energy bound state}$$

Show from Eq. (5.162) that for this potential, the s-wave phase shift satisfies

$$k \cot \delta_0(k) = \frac{1}{2}Rk^2 + O(k^4)$$

Hence conclude that the scattering length and effective range in this case are given by

$$|a| = \infty \qquad ; \; r_0 = R$$

Solution to Problem 5.14

An exact expression for $k \cot \delta_0(k)$ for the potential in Fig. 5.23 in the text was derived in Eq. (5.162)

$$k \cot \delta_0(k) = \frac{\kappa \cot(\kappa R) + k \tan(kR)}{1 - (\kappa/k) \tan(kR) \cot(\kappa R)} \qquad ; \; \kappa^2 = v_0 + k^2$$

The goal is to expand this expression through $O(k^2)$ when $v_0 R^2 = \pi^2/4$. Consider first

$$\cot(\kappa R) = \cot\left(\frac{\pi^2}{4} + k^2 R^2\right)^{1/2} = \cot\frac{\pi}{2}\left(1 + \frac{4k^2 R^2}{\pi^2}\right)^{1/2}$$

$$\approx \cot\left(\frac{\pi}{2} + \frac{k^2 R^2}{\pi}\right) = -\tan\frac{k^2 R^2}{\pi}$$

$$\approx -\frac{k^2 R^2}{\pi}$$

Since this is already of $O(k^2)$, one has

$$\kappa \cot(\kappa R) \approx -\sqrt{v_0} \frac{k^2 R^2}{\pi} = -\frac{k^2 R}{2}$$

Then also to this order, in the numerator,

$$k \tan(kR) \approx k^2 R$$

Since the numerator is now explicitly of $O(k^2)$, the denominator can be replaced by unity

$$1 - (\kappa/k) \tan(kR) \cot(\kappa R) \approx 1$$

A combination of these results gives to $O(k^2)$

$$k \cot \delta_0(k) = \frac{1}{2} R k^2$$

A comparison with the effective range formula

$$k \cot \delta_0(k) = -\frac{1}{a} + \frac{1}{2} r_0 k^2$$

allows one to conclude that the scattering length and effective range for this potential are given by

$$|a| = \infty \qquad ; \, r_0 = R$$

Problem 5.15 Derive the result for the effective range $r_0 = R$ for the potential in Prob. 5.14 from effective-range theory in Eq. (5.170).

Solution to Problem 5.15

According to Eq. (5.170), the effective range can be calculated from

$$\frac{1}{2}r_0 = \int_0^\infty dr[\phi^2(r,0) - w^2(r,0)]$$

where $\phi(r,0)$ is the extrapolation of the zero-energy solution outside the potential normalized to $\phi(r,0) = 1 - r/a$, and $w(r,0)$ is the actual solution in the potential which matches onto $\phi(r,0)$ (see Fig. 5.25 in the text). In the case in Prob. 5.14, with a bound state at zero energy, $|a| = \infty$, and $\kappa r = \sqrt{v_0}\, r = (\pi/2)(r/R)$. Thus

$$\phi(r,0) = 1$$
$$w(r,0) = \sin\left(\sqrt{v_0}\, r\right) = \sin\left(\frac{\pi}{2}\frac{r}{R}\right) \qquad ; \, 0 \le r \le R$$

There is just one-quarter wavelength inside the potential.

The expression for r_0 thus takes the form

$$\frac{1}{2}r_0 = \int_0^R dr \left[1 - \sin^2\left(\frac{\pi}{2}\frac{r}{R}\right)\right]$$
$$= R - \frac{1}{2}R = \frac{1}{2}R$$

We therefore reproduce the result

$$r_0 = R$$

Problem 5.16 Consider the hard-core–square-well potential in Fig. 5.11.

(a) Show that the condition for a zero-energy bound state is

$$\sqrt{v_0}\, b_w = \frac{\pi}{2} \qquad ; \, \text{zero-energy bound state}$$

(b) What is the appropriate modification of Eq. (5.162) in this case?

(c) Show from your result in (b) that the scattering length and effective range for this potential are given by

$$|a| = \infty \qquad ; \; r_0 = 2b + b_w$$

(d) Derive the result for the effective range in part (c) from effective-range theory in Eq. (5.170).

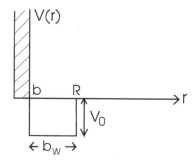

Fig. 5.11 Hard-core–square-well potential. Here $R = b + b_w$.

Solution to Problem 5.16

(a) The solution in the potential is

$$u_0(r) = A \sin \kappa (r - b) \qquad ; \; \kappa = \sqrt{v_0 + k^2}$$

For a bound state at zero energy, there must be just one-quarter wavelength inside the potential so the interior solution matches onto an exterior solution with zero slope. Therefore

$$\sqrt{v_0}(R - b) = \sqrt{v_0}\, b_w = \frac{\pi}{2}$$

(b) Since the solution inside the potential is changed to the above $u_0(r)$, Eq. (5.153) for the phase shift takes the form

$$k \cot [kR + \delta_0(k)] = \kappa \cot [\kappa(R - b)] = \kappa \cot (\kappa b_w)$$

The derivation in Eqs. (5.160)–(5.162) then proceeds with the simple replacement $\cot (\kappa R) \to \cot (\kappa b_w)$, and Eq. (5.162) becomes

$$k \cot \delta_0(k) = \frac{\kappa \cot (\kappa b_w) + k \tan (kR)}{1 - (\kappa/k) \tan (kR) \cot (\kappa b_w)} \qquad ; \; \kappa^2 = v_0 + k^2$$

(c) The analysis in the solution to Prob. 5.14 gives in the numerator of the above

$$\kappa \cot (\kappa b_w) \approx -\frac{k^2 b_w}{2}$$
$$k \tan (kR) \approx k^2 R$$

and the denominator can again be replaced by unity. Thus to $O(k^2)$

$$k \cot \delta_0(k) = -\frac{k^2 b_w}{2} + k^2 R = \frac{1}{2}(2b + b_w)k^2$$

Hence the scattering length and effective range for this potential are given by

$$|a| = \infty \qquad \qquad ; \ r_0 = 2b + b_w$$

(d) As in the solution to Prob. 5.15, we have

$$\phi(r, 0) = 1$$
$$w(r, 0) = \sin \left[\sqrt{v_0} (r - b) \right] = \sin \left[\frac{\pi}{2} \frac{(r - b)}{b_w} \right] \qquad ; \ b \leq r \leq R$$

The expression for the effective range then becomes

$$\frac{1}{2}r_0 = \int_0^R dr - \int_b^R dr \sin^2 \left[\frac{\pi}{2} \frac{(r - b)}{b_w} \right]$$
$$= R - \frac{1}{2}b_w = \frac{1}{2}(2b + b_w)$$

This reproduces the above expression for the effective range

$$r_0 = 2b + b_w$$

Problem 5.17 (a) In the WKB approximation, valid for large l, the effective potential is first modified to $\tilde{v}_{\text{eff}}(r)$ in the second of Eqs. (5.202). Show the WKB phase shift $\delta_l^{\text{WKB}}(k)$ is then obtained by adding a term of $O(1/k)$ to $\phi(r)$ in the phase of the wave functions in Eq. (5.195).

$$u_l^{(+)}(r) = \bar{a} \left\{ e^{ik[\phi(r) - \phi(r_0) + \pi/4k]} - e^{-ik[\phi(r) - \phi(r_0) + \pi/4k]} \right\}$$

(b) Show this wave function no longer vanishes at the classical turning point.

Solution to Problem 5.17

(a) In the derivation of the WKB phase shift in the text, the function $\phi(r)$ in the phase $e^{ik\phi(r)}$ is calculated up to $O(1/k)$, and the wave function $u_l^{(+)}(r)$ is then calculated under the assumption that it vanishes at the classical turning point r_0

$$u_l^{(+)}(r) = \bar{a}\left\{ e^{ik[\phi(r)-\phi(r_0)]} - e^{-ik[\phi(r)-\phi(r_0)]}\right\}$$

$$u_l^{(+)}(r_0) = 0$$

It is observed that the resulting phase shift does not vanish when the potential $v(r)$ vanishes, a necessary requirement. This situation is remedied in two steps:

- The potential is first modified from $v_{\rm eff}(r) = v(r) + l(l+1)/r^2$ to

$$\tilde{v}_{\rm eff}(r) \equiv v(r) + \frac{(l+1/2)^2}{r^2}$$

  which is valid for large l;
- It is then observed that by adding a specific term of $O(1/k)$ to ϕ, that is $\delta\phi = \pi/4k$, one obtains a phase shift that indeed vanishes for $v = 0$.

Under these modifications, the wave function is changed to

$$u_l^{(+)}(r) \rightarrow \bar{a}\left\{ e^{ik[\phi(r)-\phi(r_0)+\pi/4k]} - e^{-ik[\phi(r)-\phi(r_0)+\pi/4k]}\right\}$$

One then obtains the WKB phase shift as in the text.

(b) This new wave function no longer vanishes at the classical turning point since

$$u_l^{(+)}(r_0) = \bar{a}\left\{ e^{i\pi/4} - e^{-i\pi/4}\right\} = \sqrt{2}\, i\bar{a}$$

As anticipated, there is now some leakage into the barrier.

Problem 5.18 (a) Expand the exponential $e^{i\chi(b,k)}$ for $k \rightarrow \infty$, and show that Eqs. (5.239) then reproduce the Born approximation for the scattering amplitude when the momentum transfer lies in the transverse plane

$$f_{\rm BA}(\mathbf{q}) = -\frac{2\mu}{4\pi\hbar^2} \int d^3y\, e^{-i\mathbf{q}\cdot\mathbf{y}}\, V(y)$$

(b) Discuss the improvement achieved through the use of the full set of Eqs. (5.239).

Solution to Problem 5.18

Equations (5.239) state that

$$f(k, \theta) = \frac{k}{2\pi i} \int_0^\infty d^{(2)}b\, e^{-i\mathbf{q}\cdot\mathbf{b}} \left[e^{i\chi(b,k)} - 1 \right]$$

$$i\chi(b, k) = -\frac{i}{2k} \int_{-\infty}^\infty dz\, v\left(\sqrt{b^2 + z^2} \right)$$

Expand the exponential in powers of $v = 2\mu V/\hbar^2$. The leading term is

$$f(k, \theta) = -\frac{2\mu}{4\pi\hbar^2} \int_0^\infty d^{(2)}b \int_{-\infty}^\infty dz\, e^{-i\mathbf{q}\cdot\mathbf{b}}\, V\left(\sqrt{b^2 + z^2} \right)$$

If the momentum transfer $\mathbf{q}$ lies in the transverse plane and has no component in the z-direction, then with $\mathbf{y} = \mathbf{b} + z\mathbf{e}_z$ and $d^{(2)}b\, dz = d^3y$, one recovers

$$f_{\text{BA}}(\mathbf{q}) = -\frac{2\mu}{4\pi\hbar^2} \int d^3y\, e^{-i\mathbf{q}\cdot\mathbf{y}}\, V(y)$$

(b) Equations (5.239) express the scattering amplitude as an infinite power series in v. This set of equations sums repeated high-energy forward scatterings from the potential.

Problem 5.19 (a) The modification of Eqs. (5.256) for a circular *aperture* is

$$e^{i\chi(b,k)} = 0 \qquad ; b > R \qquad ; \text{complete absorption}$$

$$e^{i\chi(b,k)} = 1 \qquad ; b < R \qquad ; \text{no phase shift}$$

Show that one obtains the *same* diffraction pattern as in Eq. (5.259) and Fig. 5.33 in the text for a black disc.[12] This is an example of *Babinet's principle* in classical optics.[13]

(b) Can you extend this result to black discs and complementary apertures of *any* shape?

Solution to Problem 5.19

The differential elastic cross section in the eikonal limit is given in Eq. (5.246) as

$$\frac{d\sigma}{dq^2} = \frac{1}{4\pi} \left| \int_0^\infty d^{(2)}b\, e^{-i\mathbf{q}\cdot\mathbf{b}} \left[e^{i\chi(b,k)} - 1 \right] \right|^2$$

---

[12] *Hint:* Use $\int_{\text{plane}} d^{(2)}b\, e^{i\mathbf{b}\cdot\mathbf{q}} = (2\pi)^2 \delta^{(2)}(\mathbf{q})$. [Note $(2\pi)^2$ is missing from the *Hint* in the problem in the text; however, this integral *does* vanish for $\mathbf{q} \neq 0$.]

[13] See [Fetter and Walecka (2003a)].

With the assumption of Eqs. (5.256) for a black disc, this leads to the Bethe-Placzek cross section in Eq. (5.259) and Fig. 5.33 in the text.

(a) Now instead of Eqs. (5.256), assume the following complementary relations for a circular *aperture*

$$e^{i\chi(b,k)} = 0 \qquad ; b > R \qquad ; \text{complete absorption}$$
$$e^{i\chi(b,k)} = 1 \qquad ; b < R \qquad ; \text{no phase shift}$$

Write the required integral in this case as

$$\int_0^\infty d^{(2)}b\, e^{-i\mathbf{q}\cdot\mathbf{b}} \left[e^{i\chi(b,k)} - 1 \right] = \int_R^\infty d^{(2)}b\, e^{-i\mathbf{q}\cdot\mathbf{b}}[-1]$$

$$= \int_0^\infty d^{(2)}b\, e^{-i\mathbf{q}\cdot\mathbf{b}}[-1] - \int_0^R d^{(2)}b\, e^{-i\mathbf{q}\cdot\mathbf{b}}[-1]$$

The first integral vanishes away from the forward direction where $\mathbf{q} \neq 0$

$$\int_{\text{plane}} d^{(2)}b\, e^{-i\mathbf{b}\cdot\mathbf{q}} = (2\pi)^2 \delta^{(2)}(\mathbf{q})$$

$$= 0 \qquad\qquad ; \mathbf{q} \neq 0$$

Therefore, for the aperture,

$$\int_0^\infty d^{(2)}b\, e^{-i\mathbf{q}\cdot\mathbf{b}} \left[e^{i\chi(b,k)} - 1 \right] = -\int_0^R d^{(2)}b\, e^{-i\mathbf{q}\cdot\mathbf{b}}[-1] \qquad ; \mathbf{q} \neq 0$$

This is simply the negative of the amplitude for the black disc, and hence,with the absolute square, one obtains the same differential cross section and same diffraction pattern for $\mathbf{q} \neq 0$.

(b) If the aperture has an arbitrary two-dimensional shape, then a repetition of the argument in part (a) gives

$$\int_0^\infty d^{(2)}b\, e^{-i\mathbf{q}\cdot\mathbf{b}} \left[e^{i\chi(b,k)} - 1 \right] = -\int_{\text{aperture}} d^{(2)}b\, e^{-i\mathbf{q}\cdot\mathbf{b}}[-1] \qquad ; \mathbf{q} \neq 0$$

Again, this is just the negative of the result for a black disc with the same shape as the aperture, and one obtains the same diffraction pattern.

Problem 5.20 (a) Start from Eqs. (5.269)–(5.272), and prove the optical theorem in Eq. (5.273);

(b) If the target is a probability sink, then the reaction cross section defined in Eq. (5.269) is a non-negative quantity. Use the superposition axiom to then establish the unitarity inequality satisfied by the partial-wave amplitudes in Eq. (5.274).

Solution to Problem 5.20

(a) The elastic scattering amplitude is given in Eqs. (5.270) as

$$f_{\text{el}}(k, \theta) = \sum_{l=0}^{\infty} (2l + 1) \left[\frac{S_l(k) - 1}{2ik} \right] P_l(\cos \theta)$$

Use $P_l(1) = 1$, and take the imaginary part to obtain

$$\text{Im}\, f_{\text{el}}(k, 0) = \frac{1}{2k} \sum_{l=0}^{\infty} (2l + 1)\, \text{Re}\,[1 - S_l(k)]$$

The total cross section is given in Eq. (5.272) as

$$\sigma_{\text{total}} = \frac{2\pi}{k^2} \sum_{l=0}^{\infty} (2l + 1)\, \text{Re}[1 - S_l(k)]$$

Hence one obtains the optical theorem in Eq. (5.273)

$$\sigma_{\text{total}} = \frac{4\pi}{k}\, \text{Im}\, f_{\text{el}}(k, 0)$$

(b) The reaction cross section is given in Eq. (5.269) as

$$\sigma_{\text{reaction}} = \frac{\pi}{k^2} \sum_{l=0}^{\infty} (2l + 1) \left[1 - |S_l(k)|^2 \right]$$

A sufficient condition that this quantity be non-negative is given in Eq. (5.274)

$$|S_l(k)|^2 \leq 1$$

The equality holds for elastic scattering, which was proven in the text using superposition. The present problem is concerned with the inequality and reactions. We shall not give a general proof, but are here content to consider the case where only the first two s- and p-waves are absorptive so that

$$\left[1 - |S_0(k)|^2 \right] + 3 \left[1 - |S_1(k)|^2 \right] > 0$$

The goal is to show that each term individually must be positive.

The net incoming flux through a large sphere surrounding the scatterer, with the superposed wave function, was calculated in Eq. (5.117) to be

$$I_{\text{incoming}} = \frac{\hbar k}{m} \frac{\pi}{k^2} \sum_{l=0}^{\infty} (2l+1)[1-|S_l|^2]$$
$$\times [\,|\alpha|^2 + |\beta|^2 + 2\text{Re}(\alpha^\star\beta)\,P_l(\cos\theta_{k_1 k_2})]$$

The condition that this be positive is

$$I_{\text{incoming}} > 0$$

Assume only absorptive s- and p-waves so the sum goes over $l = (0,1)$. Use $P_0(x) = 1$, $P_1(x) = x$, and choose an angle so that $x = -1$. With the cancellation of positive factors, the above relations then take the form

$$|\alpha + \beta|^2 \left[1 - |S_0(k)|^2\right] + 3|\alpha - \beta|^2 \left[1 - |S_1(k)|^2\right] > 0$$

Since the ratio $|\alpha - \beta|^2/|\alpha + \beta|^2$ is arbitrary, one concludes that both $1 - |S_0(k)|^2$ and $1 - |S_0(k)|^2$ must be positive. Hence

$$|S_l(k)|^2 < 1 \qquad ; l = 0, 1$$

This is the desired result.

Problem 5.21 Show that if calculated with incoming-wave boundary conditions, the analog of Eq. (5.101) is

$$G_l^{(-)}(x, y; k) = -ikj_l(ky)h_l^{(2)}(kx) \qquad ; x > y$$
$$= -ikj_l(kx)h_l^{(2)}(ky) \qquad ; x < y$$

Solution to Problem 5.21

The calculation of the Green's function proceeds exactly as in the text up to Eq. (5.91). For $x > y$ one has

$$G_l^{(-)}(x, y; k) = \frac{1}{2\pi} \oint_{C_1} \frac{t^2 dt \, j_l(ty)h_l^{(1)}(tx)}{t^2 - (k - i\eta)^2} + \frac{1}{2\pi} \oint_{C_2} \frac{t^2 dt \, j_l(ty)h_l^{(2)}(tx)}{t^2 - (k - i\eta)^2}$$

The integrands are analytic except for the poles, whose positions are now simply the reflections through the $y = 0$ axis of those in Fig. 5.17 in the text. Evaluation with residues then gives

$$G_l^{(-)}(x,y;k) = \frac{2\pi i}{2\pi}\left[\frac{(-k)^2}{2(-k)}j_l(-ky)h_l^{(1)}(-kx) - \frac{(k)^2}{2k}j_l(ky)h_l^{(2)}(kx)\right]$$

The use of the symmetry properties in Eqs. (5.82) reduces this to

$$G_l^{(-)}(x,y;k) = -ikj_l(ky)h_l^{(2)}(kx) \qquad ; x > y$$

If $x < y$, one just reverses their role in evaluating the integral in the above

$$G_l^{(-)}(x,y;k) = -ikj_l(kx)h_l^{(2)}(ky) \qquad ; x < y.$$

Although this problem presents a nice exercise in contour integration, the result is immediately obtained by simply taking the complex conjugate of the expressions in Eqs. (5.101).

Problem 5.22 (a) Start from the integral equation satisfied by the partial-wave radial wave function $\psi_l^{(+)}(r;k)$. Show that if one defines

$$\psi_l^{(+)}(r;k) \equiv e^{i\delta_l(k)}\,\psi_l^{(s)}(r;k)$$

then $\psi_l^{(s)}(r;k)$ is *real*;

(b) Consider the partial-wave radial wave function $\psi_l^{(-)}(r;k)$ defined through the Green's function $e^{-ik|\mathbf{x}-\mathbf{y}|}/4\pi|\mathbf{x}-\mathbf{y}|$ with incoming-wave boundary conditions (see Prob. 5.21). Show

$$\psi_l^{(-)}(r;k) \equiv e^{-i\delta_l(k)}\,\psi_l^{(s)}(r;k)$$

Hence conclude that the partial-wave scattering amplitude $S_l(k)$ is obtained from

$$\psi_l^{(+)}(r;k) = e^{2i\delta_l(k)}\,\psi_l^{(-)}(r;k) = S_l(k)\,\psi_l^{(-)}(r;k)$$

Solution to Problem 5.22

(a) First, use Eq. (5.103) to show[14]

$$\frac{e^{2i\delta_l(k)} - 1}{2i} = e^{i\delta_l(k)}\sin\delta_l(k) = -ke^{i\delta_l(k)}\int_0^\infty y^2 dy\, j_l(ky)v(y)\psi_l^{(s)}(y;k)$$

$$\sin\delta_l(k) = -k\int_0^\infty y^2 dy\, j_l(ky)v(y)\psi_l^{(s)}(y;k)$$

---

[14]Here $x = |\mathbf{x}| = r$.

Now separate the Green's function in Eqs. (5.101) into two parts, where the second is explicitly real

$$G_l^{(+)}(x, y; k) = ikj_l(kx)j_l(ky) + G_l^{(s)}(x, y; k)$$
$$G_l^{(s)}(x, y; k) = -kj_l(ky)n_l(kx) \qquad ; x > y$$
$$= -kj_l(kx)n_l(ky) \qquad ; x < y$$

The integral Eq. (5.100) is then written as

$$\psi_l^{(s)}(x; k) = e^{-i\delta_l(k)}j_l(kx) - ikj_l(kx)\int_0^\infty y^2 dy \, j_l(ky)v(y)\psi_l^{(s)}(y; k)$$
$$- \int_0^\infty y^2 dy \, G_l^{(s)}(x, y; k)v(y)\psi_l^{(s)}(y; k)$$

Substitution of the previous result allows the first two terms to be re-written as

$$j_l(kx)\left[e^{-i\delta_l(k)} + i\sin\delta_l(k)\right] = j_l(kx)\cos\delta_l(k)$$

Hence $\psi_l^{(s)}(x; k)$ obeys the real integral equation

$$\psi_l^{(s)}(x; k) = j_l(kx)\cos\delta_l(k) - \int_0^\infty y^2 dy \, G_l^{(s)}(x, y; k)v(y)\psi_l^{(s)}(y; k)$$

(b) From the result in Prob. 5.21, the Green's function $G_l^{(-)}(x, y; k)$ is simply the complex conjugate of $G_l^{(+)}(x, y; k)$. The integral Eq. (5.100) implies that the solutions are similarly complex conjugates

$$\psi_l^{(-)}(x; k) = \left[\psi_l^{(+)}(x; k)\right]^\star = e^{-i\delta_l(k)}\,\psi_l^{(s)}(x; k)$$

(c) We therefore conclude that

$$\psi_l^{(+)}(x; k) = e^{2i\delta_l(k)}\,\psi_l^{(-)}(x; k) = S_l(k)\,\psi_l^{(-)}(x; k)$$

Problem 5.23 In an attempt to obtain a better description of the surface of the target than provided by the black disc, one may try a "diffuse grey disc" where the eikonal phase is defined by

$$e^{i\chi(b)} - 1 \equiv -\exp\left(-\frac{b^2\ln 2}{R^2}\right) \qquad ; \text{grey disc}$$

(a) Show that the elastic scattering cross section for the grey disc is given by

$$\left(\frac{d\sigma}{dq^2}\right)_{\text{el}} = \frac{\pi R^4}{4(\ln 2)^2} \exp\left(-\frac{q^2 R^2}{2\ln 2}\right)$$

(b) Show the integrated elastic and reaction cross sections are given by

$$\sigma_{\text{elastic}} = \frac{\pi R^2}{2\ln 2} \qquad ; \; \sigma_{\text{reaction}} = \frac{3\pi R^2}{2\ln 2}$$

(c) Compare with the results for a black disc;

(d) Compare both the grey and black disc results with the experimental cross section for p-p scattering at $E_{\text{CM}} = 53.2\,\text{GeV}$ shown in Fig. 5.34 in the text. Note $-t \equiv (\hbar c \mathbf{q})^2$.

Solution to Problem 5.23

(a) We start from Eq. (5.246) in the text

$$\frac{d\sigma}{dq^2} = \frac{1}{4\pi}\left|\int_0^\infty d^{(2)}b \; e^{-i\mathbf{q}\cdot\mathbf{b}} \left[e^{i\chi(b,k)} - 1\right]\right|^2$$

and we concentrate on the integral

$$\mathcal{I}(q, R) \equiv \int_0^\infty d^{(2)}b \; e^{-i\mathbf{q}\cdot\mathbf{b}} \left[e^{i\chi(b,k)} - 1\right]$$

An integration over the angle, and substitution of the grey-disc eikonal phase, leads to

$$\mathcal{I}(q, R) = -2\pi \int_0^\infty e^{-b^2 \ln 2/R^2} J_0(qb) \, b\,db$$

Now recall the series representation for $J_0(x)$ in Eq. (2.207)

$$J_0(x) = \sum_{l=0}^\infty \frac{(-1)^l}{(l!)^2} \left(\frac{x}{2}\right)^{2l}$$

Use this inside the last expression for $\mathcal{I}(q, R)$, and integrate term by term to obtain

$$\mathcal{I}(q, R) = -\frac{\pi R^2}{\ln 2} \sum_{l=0}^\infty \frac{(-1)^l}{l!} \left(\frac{q^2 R^2}{4\ln 2}\right)^l = -\frac{\pi R^2}{\ln 2} \exp\left(-\frac{q^2 R^2}{4\ln 2}\right)$$

This gives the desired expression for the differential cross section

$$\left(\frac{d\sigma}{dq^2}\right)_{el} = \frac{\pi R^4}{4(\ln 2)^2} \exp\left(-\frac{q^2 R^2}{2\ln 2}\right)$$

(b) The integrated elastic scattering cross section for the grey disc is obtained integrating $(d\sigma/dq^2)_{el}$ over q^2 [see Eq. (5.245) of the text]

$$\sigma_{elastic} = \int_0^\infty \left(\frac{d\sigma}{dq^2}\right)_{el} dq^2 = \frac{\pi R^2}{2\ln 2}$$

The reaction cross section is given in Eq. (5.255) and reads

$$\sigma_{reaction} = \int_0^\infty d^{(2)}b \left[1 - \left|e^{i\chi(b,k)}\right|^2\right]$$

$$= 2\pi \int_0^\infty b db \left[1 - \left(1 - e^{-b^2 \ln 2/R^2}\right)^2\right] = \frac{3\pi R^2}{2\ln 2}$$

(c) The corresponding expressions for a black disc are given in Eqs. (5.259) and (5.260) of the text. The differential cross section is

$$\left(\frac{d\sigma}{dq^2}\right)_{el} = \frac{\pi R^4}{4}\left[\frac{J_1(qR)}{qR/2}\right]^2 \qquad ; \text{ black disc}$$

and the integrated cross sections are

$$\sigma_{elastic} = \sigma_{reaction} = \pi R^2 \qquad ; \text{ black disc}$$

(d) This part asks for a comparison of these model calculations with Fig. 5.34 in the text. In Fig. 5.12 we plot $(d\sigma/dt)/(d\sigma/dt)_{t=0}$ against $-t = (\hbar c q)^2$ in GeV^2 for the case of grey and black discs. The expressions for these quantities are

$$\frac{(d\sigma/dt)}{(d\sigma/dt)_{t=0}} = \exp\left[-\frac{R^2(-t/\hbar^2 c^2)}{2\ln 2}\right] \qquad ; \text{ grey disc}$$

and

$$\frac{(d\sigma/dt)}{(d\sigma/dt)_{t=0}} = \frac{4J_1^2\left(R\sqrt{-t/\hbar^2 c^2}\right)}{R^2(-t/\hbar^2 c^2)} \qquad ; \text{ black disc}$$

The plot is produced using a transverse proton radius parameter for the grey disc of $R = 0.66\,\text{F}$ and $\hbar c = 0.1973\,\text{GeV-F}$. If we ask that the black-disc have the same slope at $t = 0$, we need to have $R_{BD} = (2/\ln 2)^{1/2}R = 1.7R = 1.12\,\text{F}$.[15]

We observe that the grey disc is able to reproduce the fall-off of the data over eight decades, but exhibits no diffraction minimum. Notice that although the black disc does exhibit a diffraction minimum, it is at the wrong location, and the second maximum of the black-disc curve is much greater that the one seen in Fig. 5.34.

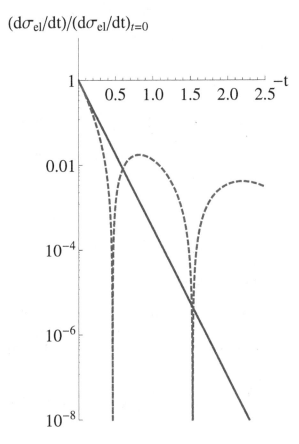

Fig. 5.12   Logarithmic plot of the ratio $(d\sigma/dt)\,/\,(d\sigma/dt)_{t=0}$ as a function of $-t = (\hbar c\mathbf{q})^2$ in GeV^2 for a grey disc (solid line) and for a black disc (dashed line) using a transverse proton radius parameter of $R \approx 0.66\,\text{F}$ for the grey disc and $R_{BD} = (2/\ln 2)^{1/2}R = 1.7R = 1.12\,\text{F}$. Compare Fig. 5.34 in the text.

---

[15]It is amusing to compare this number with the root-mean-square charge radius of a single proton measured in electron scattering of $\approx 0.8\,\text{F}$.

Problem 5.24 (a) Compute the classical attenuation of a beam going through a medium of scatterers with elementary cross sections σ_t and density distribution $\rho(x)$. Justify the identification of the classical transmission with $|e^{i\chi}|^2$, and hence show that the high-energy phase and *optical potential* for a composite system take the form

$$i\chi_{\text{opt}}(b) = -\frac{1}{2}\sigma_t \int_{-\infty}^{\infty} dz\, \rho(\sqrt{b^2 + z^2})$$

$$v_{\text{opt}}(x) = -ik\sigma_t\rho(x) \qquad \text{; optical potential}$$

(b) Use the optical theorem, and restore the full forward scattering amplitude, to obtain the following extended expression for the optical potential[16]

$$v_{\text{opt}}(x) = -4\pi f_{\text{el}}(k,0)\rho(x) \qquad \text{; optical potential}$$

When is the result in (a) then applicable?[17]

Solution to Problem 5.24

(a) Consider the one-dimensional problem of a beam of intensity I_0 incident on a slab with scatterer number density $\rho(x)$, each with cross section σ_t. The decrease in intensity in traversing a slice of thickness dx and unit transverse area is

$$dI = -I\sigma_t\rho(x)dx$$

This expression is integrated through the slab to give the classical transmission

$$\frac{I}{I_0} = \exp\left[-\sigma_t \int_{-\infty}^{\infty} \rho(x)dx\right]$$

It follows from the discussion in section 5.8.2, that at high energy in quantum mechanics with a complex potential $v(r)$, the probability of traversing the medium at an impact parameter b is given by the straight-line eikonal result

$$|e^{i\chi(b,k)}|^2 = \left|\exp\left[-\frac{i}{2k}\int_{-\infty}^{\infty} v(\sqrt{b^2 + z^2})dz\right]\right|^2$$

---

[16] See, for example, [Walecka (2004)].
[17] *Hint:* Contemplate Eq. (5.274) and the second of Eqs. (5.270).

A comparison of these two expressions allows one to define an effective absorptive optical potential for the medium, where now $x = |\mathbf{x}| = r$,

$$v_{\text{opt}}(x) = -ik\sigma_t \rho(x) \qquad \text{; optical potential}$$

Then

$$|e^{i\chi_{\text{opt}}(b,k)}|^2 = \left|\exp\left[-\frac{\sigma_t}{2}\int_{-\infty}^{\infty}\rho(\sqrt{b^2+z^2})dz\right]\right|^2$$

$$= \exp\left[-\sigma_t\int_{-\infty}^{\infty}\rho(\sqrt{b^2+z^2})dz\right]$$

which coincides with the classical transmission.

(b) The optical theorem in Eq. (5.273) relates the total cross section to the imaginary part of the forward scattering amplitude

$$\sigma_t = \frac{4\pi}{k}\text{Im}\,f_{\text{el}}(k,0)$$

Thus the optical potential in part (a) can be written

$$v_{\text{opt}}(x) = -4\pi i\,\text{Im}\,f_{\text{el}}(k,0)\,\rho(x)$$

An improvement in this expression (here stated without proof), which takes into account refraction as well as absorption, is to replace $i\,\text{Im}\,f_{\text{el}}(k,0) \to f_{\text{el}}(k,0)$[18]

$$v_{\text{opt}}(x) = -4\pi f_{\text{el}}(k,0)\,\rho(x)$$

If there is complete absorption so that for the contributing partial waves $|S_l(k)| = 0$ and $f_{\text{el}}(k,0)$ is imaginary [see Eqs. (5.274) and (5.270)], then

$$f_{\text{el}}(k,0) = \text{Re}\,f_{\text{el}}(k,0) + i\,\text{Im}\,f_{\text{el}}(k,0) \approx i\,\text{Im}\,f_{\text{el}}(k,0)$$

and the two expressions for the optical potential coincide.

---

[18]See, for example, [Walecka (2004)].

Chapter 6

Time-Dependent Perturbation Theory

Problem 6.1 (a) Make a good numerical calculation of the second integral in Eqs. (6.39) to evaluate and plot $K_0(\alpha)$;

(b) Take the derivative of your result in (a) to evaluate and plot $K_1(\alpha)$;

(c) Compare your numerical results with the two asymptotic expressions in Eqs. (6.41).

Solution to Problem 6.1

(a) With a change of variable, we may write the second integral of Eqs. (6.39) as

$$K_0(\alpha) = \int_0^\infty \frac{\cos u}{\sqrt{\alpha^2 + u^2}} du$$

As it stands, the integral is not in a form that is convenient for a numerical calculation because of the oscillatory behavior of the integrand. We first do some analysis to transform the integral into a form more readily adopted to numerical work. The integral may be re-cast in the form

$$K_0(\alpha) = \int_0^\infty \frac{e^{iu}}{2\sqrt{\alpha^2 + u^2}} du + \int_0^\infty \frac{e^{-iu}}{2\sqrt{\alpha^2 + u^2}} du \equiv \mathcal{I}_1 + \mathcal{I}_2$$

and we now work separately on the two integrals $\mathcal{I}_1$ and $\mathcal{I}_2$.

In the case of the first integral it is convenient to consider the contour of integration shown in Fig. 6.1, of which $\mathcal{I}_1$ is the part on the real axis. Let us define

$$f(u) \equiv \frac{e^{iu}}{2\sqrt{\alpha^2 + u^2}}$$

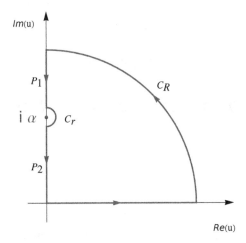

Fig. 6.1   Contour integral for $\mathcal{I}_1$ in the complex u plane.

With the use of the residue theorem we have

$$\mathcal{I}_1 = \lim_{R \to \infty} \left[-\int_{C_R} f(u)du - \int_{P_1} f(u)du - \int_{C_r} f(u)du - \int_{P_2} f(u)du \right]$$

where the integrals are evaluated moving anticlockwise. It is easy to see that

$$\lim_{R \to \infty} \int_{C_R} f(u)du = 0 \qquad ; \quad \lim_{r \to 0} \int_{C_r} f(u)du = 0$$

Similarly, we have

$$\lim_{R \to \infty} \int_{P_1} f(u)du = -\int_\alpha^\infty \frac{e^{-v}}{2\sqrt{v^2 - \alpha^2}}dv$$

and

$$-\lim_{R \to \infty} \int_{P_2} f(u)du = \int_0^{i\alpha} f(u)du$$

Let us now come to the second integral; we define

$$g(u) \equiv \frac{e^{-iu}}{2\sqrt{\alpha^2 + u^2}} = f(-u)$$

and consider the contour of integration shown in Fig. 6.2. From the residue theorem we have

$$\mathcal{I}_2 = \lim_{R \to \infty} \left[-\int_{C_R'} g(u)du - \int_{P_1'} g(u)du - \int_{C_r'} g(u)du - \int_{P_2'} g(u)du \right]$$

where the integrals are evaluated moving clockwise.

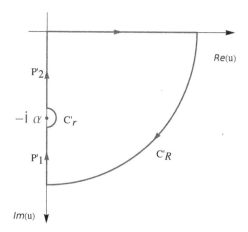

Fig. 6.2 Contour integral for $\mathcal{I}_2$ in the complex u plane.

Once again we have

$$\lim_{R \to \infty} \int_{C_R'} g(u)du = 0 \qquad ; \lim_{r \to 0} \int_{C_r'} g(u)du = 0$$

Similarly we have

$$\lim_{R \to \infty} \int_{P_1'} g(u)du = -\int_{\alpha}^{\infty} \frac{e^{-v}}{2\sqrt{v^2 - \alpha^2}} dv$$

and

$$-\lim_{R \to \infty} \int_{P_2'} g(u)du = -\int_{-i\alpha}^{0} f(-u)du = -\int_{0}^{i\alpha} f(u)du$$

where we have used the property $g(u) = f(-u)$.

With the observation that the integrals over P_2 and P_2' cancel each other, we have

$$K_0(\alpha) = \mathcal{I}_1 + \mathcal{I}_2 = \int_{\alpha}^{\infty} \frac{e^{-v}}{\sqrt{v^2 - \alpha^2}} dv$$

This is now in a form suitable for a numerical calculation, since the integrand is a monotonous function. The function $K_0(\alpha)$ is plotted in Fig. 6.3.

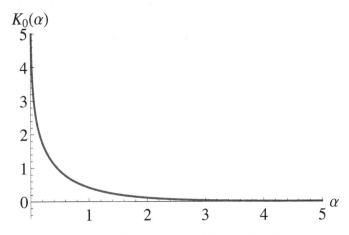

Fig. 6.3   Numerical evaluation of the function $K_0(\alpha)$.

(b) We first write the result of part (a) as

$$K_0(\alpha) = \lim_{\eta \to 0^+} \int_{\alpha+\eta}^{\infty} \frac{e^{-v}}{\sqrt{v^2 - \alpha^2}} dv$$

and then evaluate

$$\frac{dK_0(\alpha)}{d\alpha} = \lim_{\eta \to 0^+} \left[-\frac{e^{-\alpha-\eta}}{\sqrt{\eta(2\alpha + \eta)}} + \int_{\alpha+\eta}^{\infty} \frac{\alpha e^{-v}}{(v^2 - \alpha^2)^{3/2}} dv \right]$$

where each term separately diverges when $\eta \to 0^+$, whereas their sum is finite.[1] A numerical calculation of the quantity $dK_0/d\alpha$ is plotted as the solid line in Fig. 6.4.

(c) In Fig. 6.4 we also compare $dK_0(\alpha)/d\alpha$ with the asymptotic expressions in Eq. (6.41) of the text. Notice how well they work in the respective asymptotic regimes.

Problem 6.2[2] A hydrogen atom in its ground state is placed between the plates of a condenser. A voltage pulse is applied to the condenser so as

---

[1] Clearly, in a numerical calculation one must use a small but finite η.
[2] See [Schiff (1968)], Prob. 8.1.

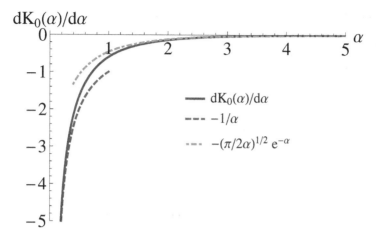

Fig. 6.4 Numerical calculation of the quantity $dK_0(\alpha)/d\alpha$ (solid line) and its asymptotic behaviors for $\alpha \to 0$ (dashed line) and $\alpha \to \infty$ (dashed-dot line) in Eqs. (6.41).

to produce a homogeneous electric field that has a time dependence

$$\mathcal{E}(t) = 0 \qquad\quad ; t < 0$$
$$\mathcal{E}(t) = \mathcal{E}_0\, e^{-t/\tau} \quad ; t > 0$$

(a) Find the first-order probability that the atom is in the $2s$-state $|200\rangle$ after a long time;

(b) What is the corresponding probability that it is one of the $2p$-states?

Solution to Problem 6.2

(a) Let the field $\mathcal{E}(t)$ lie along the z-direction so that

$$\mathcal{E}(t) = \mathcal{E}_0 e^{-t/\tau}\, \mathbf{e}_z$$

The potential and perturbation are then

$$\Phi(t) = -\mathcal{E}_0 z e^{-t/\tau}$$
$$H'(t) = -e\mathcal{E}_0 z e^{-t/\tau} = -e\mathcal{E}_0 \left(\frac{4\pi}{3}\right)^{1/2} r Y_{10}(\theta, \phi) e^{-t/\tau}$$

Write this as

$$H'(t) = H' e^{-t/\tau}$$
$$H' = -e\mathcal{E}_0 \left(\frac{4\pi}{3}\right)^{1/2} r Y_{10}(\theta, \phi)$$

The first-order transition probability follows from Eq. (6.26) as

$$P_{fi}(\infty, 0) = \left| -\frac{i}{\hbar} \int_0^\infty dt \, \langle \phi_f | H'(t) | \phi_i \rangle e^{i(E_f - E_i)t/\hbar} \right|^2$$

The time integral is immediately performed to give

$$P_{fi}(\infty, 0) = \left| \frac{i \langle \phi_f | H' | \phi_i \rangle}{i\Delta E - \hbar/\tau} \right|^2 \qquad ; \, \Delta E \equiv E_f - E_i$$

$$= \frac{|\langle \phi_f | H' | \phi_i \rangle|^2}{(\Delta E)^2 + (\hbar/\tau)^2}$$

Consider the contribution of the angular part of the remaining matrix element

$$\frac{4\pi}{3} \left| \int d\Omega \, Y^\star_{l_f, m_f}(\theta, \phi) Y_{10}(\theta, \phi) Y_{00} \right|^2 = \frac{1}{3} \delta_{l_f,1} \delta_{m_f,0}$$

It follows that of the four excited $|200\rangle, |21m\rangle$ states, *only the $|210\rangle$ is excited from the ground $|100\rangle$ during the discharge.*

(b) The remaining required radial integral is evaluated in Eq. (7.166)

$$\int_0^\infty r^2 dr \, R_{2p}(r) r R_{1s}(r) = \frac{1}{\sqrt{6}} \frac{2^8}{3^4} \frac{a_0}{Z}$$

Hence the probability for the $|100\rangle \to |210\rangle$ transition in hydrogen is[3]

$$P_{2p0 \leftarrow 1s}(\infty, 0) = \frac{2^{15}}{3^{10}} \frac{(e\mathcal{E}_0 a_0)^2}{(\Delta E)^2 + (\hbar/\tau)^2} \qquad ; \, \Delta E = E_{2p}^0 - E_{1s}^0$$

Problem 6.3 Use the half-width of the central peak in Fig. 6.8 in the text as a measure $\Delta\omega$ of the spread in frequencies corresponding to the time interval $T \equiv \Delta t$. Hence derive the *effective* uncertainty relation

$$\Delta E \Delta t \sim h \qquad ; \text{ effective uncertainty relation}$$

Note, however, that this result is obtained in a significantly different fashion than the uncertainty principle in Eq. (3.221).

---

[3]For hydrogen $Z = 1$.

Solution to Problem 6.3

The full width Γ at half-maximum of the central peak in Fig. 6.8 in the text, in the energy $\hbar\omega$, is

$$\Gamma \sim \frac{2\pi\hbar}{T}$$

Suppose one takes a time interval $\Delta t \equiv T$ and an energy interval $\Delta E \equiv \Gamma$. Then there is an *effective* uncertainty relation

$$\Delta E \Delta t \sim \frac{2\pi\hbar}{T} T \sim h$$

Problem 6.4 (a) Show the cross section for exciting the H-atom from the $1s$ to the $2s$ state through inelastic charged-lepton $l^{\pm}$ scattering is given to leading order in $\alpha = e^2/\hbar c$ by

$$\left(\frac{d\sigma}{d\Omega}\right)_{2s \leftarrow 1s} = \frac{\alpha^2}{q^4}\left(\frac{m_l c}{\hbar}\right)^2 \frac{k_f}{k_i} \frac{128(qa_0)^4}{[(qa_0)^2 + 9/4]^6}$$

Here $a_0 = \hbar^2/m_e e^2$ is the electron Bohr radius.

(b) Show the cross section for the $1s \to 2p$ transition is

$$\left(\frac{d\sigma}{d\Omega}\right)_{2p \leftarrow 1s} = \frac{\alpha^2}{q^4}\left(\frac{m_l c}{\hbar}\right)^2 \frac{k_f}{k_i} \frac{288(qa_0)^2}{[(qa_0)^2 + 9/4]^6}$$

Solution to Problem 6.4

The cross section for inelastic scattering is given in Eq. (6.116) as

$$\frac{d\sigma_{fi}}{d\Omega} = \frac{4}{q^4}\left(\frac{e^2}{\hbar c}\right)^2 \left(\frac{m_l c}{\hbar}\right)^2 \left(\frac{k_f}{k_i}\right) |\mathcal{F}_{fi}(\mathbf{q})|^2$$

where $\mathcal{F}_{fi}(\mathbf{q})$ is the inelastic form factor given in Eq. (6.129) as

$$\mathcal{F}_{nlm,1s}(\mathbf{q}) = \delta_{m0}(-i)^l (2l+1)^{1/2} \int_0^\infty r^2 dr\, R_{nl}(r) j_l(qr) R_{1s}(r)$$

The required radial wave functions are given in Prob. 2.22

$$R_{10}(r) = \left(\frac{Z}{a_0}\right)^{3/2} 2e^{-Zr/a_0}$$

$$R_{20}(r) = \left(\frac{Z}{2a_0}\right)^{3/2} \left(2 - \frac{Zr}{a_0}\right) e^{-Zr/2a_0}$$

$$R_{21}(r) = \left(\frac{Z}{2a_0}\right)^{3/2} \frac{Zr}{a_0\sqrt{3}} e^{-Zr/2a_0}$$

(a) For the $1s \to 2s$ transition we need the radial matrix element

$$\int_0^\infty r^2 dr\, R_{2s}(r) j_0(qr) R_{1s}(r) = \left(\frac{Z}{a_0}\right)^3 \frac{1}{\sqrt{2}} \times$$
$$\int_0^\infty r^2 dr \left(2 - \frac{Zr}{a_0}\right) j_0(qr) e^{-3Zr/2a_0}$$

To do the integrals, consider

$$\int_0^\infty t^2 dt\, j_0(qt) e^{-at} = \int_0^\infty t^2 dt\, \frac{\sin qt}{qt} e^{-at} = -\frac{\partial}{\partial a} \int_0^\infty t dt\, \frac{\sin qt}{qt} e^{-at}$$
$$= -\frac{\partial}{\partial a} \left\{ \frac{1}{2iq} \int_0^\infty dt\, e^{-at} \left[e^{iqt} - e^{-iqt}\right] \right\}$$
$$= \frac{\partial}{\partial a} \frac{1}{2iq} \left[\frac{1}{iq-a} + \frac{1}{iq+a} \right] = -\frac{\partial}{\partial a} \frac{1}{q^2+a^2}$$
$$= \frac{2a}{(q^2+a^2)^2}$$

Thus we have

$$\int_0^\infty t^2 dt\, j_0(qt) e^{-at} = \frac{2a}{(q^2+a^2)^2}$$
$$\int_0^\infty t^3 dt\, j_0(qt) e^{-at} = \frac{6a^2 - 2q^2}{(q^2+a^2)^3}$$

where the second result comes from taking $-\partial/\partial a$ on the first.

The required radial matrix element is therefore evaluated as

$$\int_0^\infty r^2 dr\, R_{2s}(r) j_0(qr) R_{1s}(r) = \left(\frac{Z}{a_0}\right)^4 \frac{1}{\sqrt{2}}$$

$$\times \left[\frac{6}{[q^2 + (3Z/2a_0)^2]^2} - \frac{6(3Z/2a_0)^2 - 2q^2}{[q^2 + (3Z/2a_0)^2]^3}\right]$$

$$= \frac{4\sqrt{2}\,(qa_0/Z)^2}{[(qa_0/Z)^2 + 9/4]^3}$$

The $1s \to 2s$ inelastic cross section for hydrogen ($Z = 1$) is thus given by

$$\left(\frac{d\sigma}{d\Omega}\right)_{2s\leftarrow 1s} = \frac{\alpha^2}{q^4}\left(\frac{m_l c}{\hbar}\right)^2 \frac{k_f}{k_i} \frac{2^7(qa_0)^4}{[(qa_0)^2 + 9/4]^6}$$

(b) For the $1s \to 2p$ transition we need the radial matrix element

$$\int_0^\infty r^2 dr\, R_{2p}(r) j_1(qr) R_{1s}(r) = \left(\frac{Z}{a_0}\right)^4 \left(\frac{1}{6}\right)^{1/2} \int_0^\infty r^3 dr\, j_1(qr) e^{-3Zr/2a_0}$$

With the use of $dj_0(u)/du = -j_1(u)$, the required radial integral is computed from the previous result as

$$\int_0^\infty t^3 dt\, j_1(qt) e^{-at} = -\frac{\partial}{\partial q}\int_0^\infty t^2 dt\, j_0(qt) e^{-at}$$

$$= \frac{8aq}{(q^2 + a^2)^3}$$

Hence

$$\int_0^\infty r^2 dr\, R_{2p}(r) j_1(qr) R_{1s}(r) = \left(\frac{Z}{a_0}\right)^4 \left(\frac{1}{6}\right)^{1/2} \frac{8(3Z/2a_0)q}{[q^2 + (3Z/2a_0)^2]^3}$$

$$= \left(\frac{1}{6}\right)^{1/2} \frac{12(qa_0/Z)}{[(qa_0/Z)^2 + 9/4]^3}$$

The $1s \to 2p$ inelastic cross section for hydrogen with $Z = 1$ is thus

$$\left(\frac{d\sigma}{d\Omega}\right)_{2p\leftarrow 1s} = \frac{\alpha^2}{q^4}\left(\frac{m_l c}{\hbar}\right)^2 \frac{k_f}{k_i} \frac{2^5 3^2(qa_0)^2}{[(qa_0)^2 + 9/4]^6}$$

Both of these results for the cross sections agree with those given in the statement of the problem.

Problem 6.5 Make a good numerical calculation and plot of the absolute square of the Coulomb inelastic atomic form factor in Eq. (6.129) for the following transitions:$1s \to 3s, 1s \to 3p$, and $1s \to 3d$. Compare and discuss.

Solution to Problem 6.5

The Coulomb inelastic atomic form factor in Eq. (6.129) reads

$$\mathcal{F}_{nlm,1s}(\mathbf{q}) = \delta_{m0}(-i)^l(2l+1)^{1/2}\int_0^\infty r^2 dr\, R_{nl}(r) j_l(qr) R_{1s}(r)$$

To solve this problem we need the explicit form of the radial wave function corresponding to the different quantum numbers. An extension of the results in Prob. 2.22 for the radial wave functions $R_{nl}(r)$ for hydrogen gives

$$R_{10}(r) = \left(\frac{1}{a_0}\right)^{3/2} 2e^{-r/a_0}$$

$$R_{30}(r) = \left(\frac{1}{3a_0}\right)^{3/2}\left(2 - \frac{4r}{3a_0} + \frac{4r^2}{27a_0^2}\right)e^{-r/3a_0}$$

$$R_{31}(r) = \left(\frac{2}{27a_0}\right)^{3/2}\frac{r}{a_0}\left(6 - \frac{r}{a_0}\right)e^{-r/3a_0}$$

$$R_{32}(r) = \left(\frac{2}{27a_0}\right)^{3/2}\frac{1}{\sqrt{5}}\frac{r^2}{a_0^2}e^{-r/3a_0}$$

where $a_0 = \hbar^2/m_e e^2$.

The required spherical Bessel functions are

$$j_0(z) = \frac{\sin z}{z}$$

$$j_1(z) = \frac{\sin z}{z^2} - \frac{\cos z}{z}$$

$$j_2(z) = \left(\frac{3}{z^3} - \frac{1}{z}\right)\sin z - \frac{3}{z^2}\cos z$$

The decreasing exponentials imply the Coulomb form factors are nice, convergent integrals, and in Fig. 6.5 we display a numerical calculation of the absolute squares of the Coulomb inelastic form factor for the transitions $1s \to 3s, 1s \to 3p$, and $1s \to 3d$, using the formulae obtained above.

Notice the characteristic q^l behavior of the inelastic form factors for the $1s \to 3p$ and $1s \to 3d$ transitions arising from the long-wavelength behavior of the spherical Bessel functions.[4]

$$j_l(z) = \frac{z^l}{(2l+1)!!} \qquad ; z \to 0$$

The orthogonality of the radial wave functions for the $1s \to 3s$ transition moves this behavior to the next order in the expansion of $j_0(z)$.

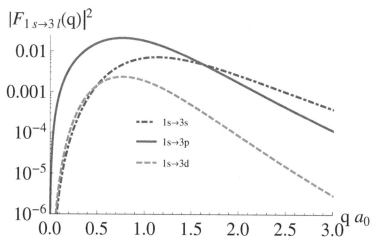

Fig. 6.5   Logarithmic plot of the absolute squares of the Coulomb inelastic form factors in hydrogen for the transitions $1s \to 3s$, $1s \to 3p$, and $1s \to 3d$.

****

(*Aside*) Although not asked for in the problem, the required inelastic form factors can be obtained analytically. The results are

$$\mathcal{F}_{300,1s}(\mathbf{q}) = \frac{3^3\sqrt{3}}{2^8} q^2 a_0^2 \frac{(1 + 27q^2 a_0^2/16)}{(1 + 9q^2 a_0^2/16)^4}$$

$$\mathcal{F}_{310,1s}(\mathbf{q}) = -2i\sqrt{\frac{2}{3}} \frac{1}{qa_0} \mathcal{F}_{300,1s}(\mathbf{q})$$

$$\mathcal{F}_{320,1s}(\mathbf{q}) = -\frac{\sqrt{2}}{1 + 27q^2 a_0^2/16} \mathcal{F}_{300,1s}(\mathbf{q})$$

where the form factors corresponding to the transitions $1s \to 3p$ and $1s \to 3d$ are expressed in terms of the form factor corresponding to the transition

---

[4]In electron scattering from nuclei this behavior is often used both to identify the multipolarity of transitions and to bring out high-spin states [Walecka (2001)].

$1s \to 3s$.

The asymptotic behavior of the form factors can then be obtained analytically from these expressions. For $q \to 0$ one has

$$|\mathcal{F}_{300,1s}(\mathbf{q})|^2 \approx \frac{3^7}{2^{16}}(qa_0)^4 \qquad ; \; qa_0 \to 0$$

$$|\mathcal{F}_{310,1s}(\mathbf{q})|^2 \approx \frac{3^6}{2^{13}}(qa_0)^2$$

$$|\mathcal{F}_{320,1s}(\mathbf{q})|^2 \approx \frac{3^7}{2^{15}}(qa_0)^4$$

For $q \to \infty$, they behave as

$$|\mathcal{F}_{300,1s}(\mathbf{q})|^2 \approx \frac{2^8}{3^3}\frac{1}{(qa_0)^8} \qquad ; \; qa_0 \to \infty$$

$$|\mathcal{F}_{310,1s}(\mathbf{q})|^2 \approx \frac{2^{11}}{3^4}\frac{1}{(qa_0)^{10}}$$

$$|\mathcal{F}_{320,1s}(\mathbf{q})|^2 \approx \frac{2^{17}}{3^9}\frac{1}{(qa_0)^{12}}$$

****

Problem 6.6 Show the total inelastic cross sections in Prob. 6.4, at high incident energy where $ka_0 \to \infty$, become[5]

$$\sigma_{2s\leftarrow1s} = \frac{128\pi}{5k^2}\left(\frac{m_l}{m_e}\right)^2\left(\frac{2}{3}\right)^{10} \qquad ; \; ka_0 \to \infty$$

$$\sigma_{2p\leftarrow1s} = \frac{576\pi}{k^2}\left(\frac{m_l}{m_e}\right)^2\left(\frac{2}{3}\right)^{12}\ln(ka_0)$$

Hint: Convert from $\int \sin\theta \, d\theta$ to $\int q\,dq$.

Solution to Problem 6.6

We are interested in the high-energy limit, where $ka_0 \to \infty$, of the integrated cross section in Prob. 6.4

$$\sigma = \int d\Omega \left(\frac{d\sigma}{d\Omega}\right)$$

---

[5]Compare [Schiff (1968)], Prob. 9.14.

First, change to an integration over the momentum transfer

$$q^2 = (\mathbf{k}_f - \mathbf{k}_i)^2 = k_f^2 + k_i^2 - 2k_f k_i \cos\theta$$
$$dq^2 = -2k_f k_i d\cos\theta = 2k_f k_i \sin\theta \, d\theta$$

Since $\int d\phi = 2\pi$, this gives

$$\sigma \to \frac{\pi}{k^2} \int dq^2 \left(\frac{d\sigma}{d\Omega}\right)$$

The integrated cross sections in Prob. 6.4 then require the evaluation of

$$\int d\Omega \, \frac{\alpha^2}{q^4} \left(\frac{m_l c}{\hbar}\right)^2 \frac{k_f}{k_i} \to \frac{\pi\alpha^2}{k^2} \left(\frac{m_l c}{\hbar}\right)^2 \int \frac{dq^2}{q^4}$$

With the definition $u \equiv (qa_0)^2$, and the use of $a_0 = \hbar/\alpha m_e c$, one has

$$\int d\Omega \, \frac{\alpha^2}{q^4} \left(\frac{m_l c}{\hbar}\right)^2 \frac{k_f}{k_i} \to \frac{\pi\alpha^2}{k^2} \left(\frac{m_l c}{\hbar}\right)^2 a_0^2 \int \frac{du}{u^2}$$

$$= \frac{\pi}{k^2} \left(\frac{m_l}{m_e}\right)^2 \int \frac{du}{u^2} \qquad ; \ u \equiv (qa_0)^2$$

We still need the limits of integration. The maximum value of the momentum transfer is given by

$$q_{\max}^2 = (k_f + k_i)^2$$

Therefore

$$u_{\max} \to 4(ka_0)^2 \to \infty$$

The minimum value of q^2 is

$$q_{\min}^2 = (k_f - k_i)^2$$

This is determined by energy conservation

$$\frac{\hbar^2 k_i^2}{2m_l} - \frac{\hbar^2 k_f^2}{2m_l} = \Delta E \qquad ; \ \Delta E = E_2^0 - E_1^0$$

$$k_i^2 - k_f^2 = \frac{2m_l \Delta E}{\hbar^2}$$

It follows that

$$k_i - k_f = \frac{2m_l \Delta E}{\hbar^2 (k_i + k_f)} \to \frac{m_l \Delta E}{\hbar^2 k}$$

Therefore

$$u_{\min} \propto \frac{1}{(ka_0)^2} \to 0$$

(a) For the asymptotic form of the $1s \to 2s$ cross section, one then has

$$\sigma_{2s \leftarrow 1s} \to \frac{\pi}{k^2} \left(\frac{m_l}{m_e}\right)^2 2^7 \int_0^\infty \frac{du}{u^2} \frac{u^2}{[u+9/4]^6}$$

$$= \frac{128\pi}{5k^2} \left(\frac{m_l}{m_e}\right)^2 \left(\frac{2}{3}\right)^{10} \qquad ; ka_0 \to \infty$$

(b) The $1s \to 2p$ cross section is a little more subtle. Here the integral over u has a logarithmic divergence at the lower limit, and this produces the asymptotic behavior

$$\sigma_{2p \leftarrow 1s} \to \frac{\pi}{k^2} \left(\frac{m_l}{m_e}\right)^2 2^5 3^2 \int_{u_{\min}}^\infty \frac{du}{u^2} \frac{u}{[u+9/4]^6}$$

$$\to \frac{\pi}{k^2} \left(\frac{m_l}{m_e}\right)^2 2^5 3^2 \left(\frac{2}{3}\right)^{12} \int_{u_{\min}} \frac{du}{u}$$

where[6]

$$\int_{u_{\min}} \frac{du}{u} \to -\ln \frac{1}{(ka_0)^2} = 2\ln(ka_0)$$

Thus the asymptotic form of the $1s \to 2p$ cross section is given by

$$\sigma_{2p \leftarrow 1s} = \frac{576\pi}{k^2} \left(\frac{m_l}{m_e}\right)^2 \left(\frac{2}{3}\right)^{12} \ln(ka_0) \qquad ; ka_0 \to \infty$$

Both of these results agree with those stated in the problem.

Problem 6.7 A development analogous to that in the text can be used to describe *direct nuclear reactions*. Assume free plane-wave states for the projectile nucleon and single-nucleon states $\psi_{nlm}(\mathbf{r})$ for the target nucleons, of the type shown in Fig. 2.11 in the text and Fig. 2.16. Assume the projectile and target nucleons interact through a two-nucleon potential of the form $V(|\mathbf{r}_p - \mathbf{r}|)$. Use the Golden Rule to show the cross section for the target transition $|n00\rangle \to |n'lm\rangle$ is given in plane-wave Born approximation

---

[6] Any constant multiplicative factors in $u_{\min}$ do not affect this asymptotic behavior.

by

$$\left(\frac{d\sigma}{d\Omega}\right)_{n'lm \leftarrow n00} = \frac{k_f}{k_i}\left|\frac{2m_p}{4\pi\hbar^2}\tilde{V}(q)\right|^2 |F_{n'lm,n00}(\mathbf{q})|^2 \quad ; \text{plane-wave B.A.}$$

$$F_{n'lm,n00}(\mathbf{q}) = \delta_{m0}(-i)^l(2l+1)^{1/2}\int_0^\infty r^2 dr\, R_{n'l}(r)j_l(qr)R_{n0}(r)$$

$$; \text{inelastic nuclear f.f.}$$

Here m_p is the nucleon mass, and $\tilde{V}(q)$ is the Fourier transform of the two-body potential in Eqs. (5.43).

Solution to Problem 6.7

This solution closely follows the discussion in section 6.2.2.3. The initial and final states are given as in Eqs. (6.100)[7]

$$\phi_i = \psi_{n00}(\mathbf{r})\frac{1}{\sqrt{\Omega}}e^{i\mathbf{k}_i\cdot\mathbf{r}_p} \quad ; E_i = \varepsilon_i^0 + \frac{\hbar^2 k_i^2}{2m_p}$$

$$\phi_f = \psi_{n'lm}(\mathbf{r})\frac{1}{\sqrt{\Omega}}e^{i\mathbf{k}_f\cdot\mathbf{r}_p} \quad ; E_f = \varepsilon_f^0 + \frac{\hbar^2 k_f^2}{2m_p}$$

The transition matrix element of the interaction hamiltonian is then calculated as in Eqs. (6.102)–(6.103)

$$\langle\phi_f|H_1|\phi_i\rangle = \frac{1}{\Omega}\int d^3r\int d^3r_p\, e^{-i\mathbf{k}_f\cdot\mathbf{r}_p}\,\psi_f^\star(\mathbf{r})V(|\mathbf{r_p}-\mathbf{r}|)\psi_i(\mathbf{r})\,e^{i\mathbf{k}_i\cdot\mathbf{r}_p}$$

$$= \frac{\tilde{V}(q)}{\Omega}\int d^3r\, e^{-i\mathbf{q}\cdot\mathbf{r}}\psi_f^\star(\mathbf{r})\psi_i(\mathbf{r}) \quad ; \mathbf{q} = \mathbf{k}_f - \mathbf{k}_i$$

where $\tilde{V}(q)$ is the Fourier transform of the two-body potential in Eqs. (5.43). The preceding expression is written in terms of the inelastic form factor in Eqs. (6.126) and (6.129) as

$$\langle\phi_f|H_1|\phi_i\rangle = \frac{\tilde{V}(q)}{\Omega}F_{n'lm,n00}(\mathbf{q})$$

with $F_{n'lm,n00}(\mathbf{q})$ given above. The density of final states and incident flux are calculated exactly as in Eqs. (6.111) and (6.113)

$$\frac{\partial n_f}{\partial E_f} = \frac{\Omega}{(2\pi)^3}k_f^2 d\Omega_f\left(\frac{m_p}{\hbar^2 k_f}\right)$$

$$I_{\text{inc}} = \frac{1}{\Omega}\left(\frac{\hbar k_i}{m_p}\right)$$

---

[7]The projectile and target nucleons are here assumed to be distinguishable.

The cross section follows as in Eq. (6.115)

$$d\sigma_{fi} = \frac{2\pi}{\hbar} \left| \frac{\tilde{V}(q)}{\Omega} F_{fi}(\mathbf{q}) \right|^2 \left[\frac{\Omega}{(2\pi)^3} k_f^2 d\Omega_f \left(\frac{m_p}{\hbar^2 k_f} \right) \right] \left[\frac{1}{\hbar k_i / \Omega m_p} \right]$$

A combination of factors then gives

$$\left(\frac{d\sigma}{d\Omega} \right)_{n'lm \leftarrow n00} = \frac{k_f}{k_i} \left| \frac{2m_p}{4\pi\hbar^2} \tilde{V}(q) \right|^2 |F_{n'lm,\, n00}(\mathbf{q})|^2$$

Here the inelastic form factor for transitions from a 1s-state is given by Eq. (6.129) as

$$F_{nlm,1s}(\mathbf{q}) = \delta_{m0}(-i)^l (2l+1)^{1/2} \int_0^\infty r^2 dr\, R_{nl}(r) j_l(qr) R_{1s}(r)$$

Problem 6.8 An improvement over the result in Prob. 6.7 is to let the initial and final projectile wave functions be distorted by the same nuclear potential seen by the target nucleon. Use the analysis in Eqs. (5.70)–(5.75) to show that the proper recipe for obtaining *distorted-wave Born approximation* is to make the following replacement in the projectile matrix elements of the two-body potential

$$\langle \mathbf{k}_f | V | \mathbf{k}_i \rangle \rightarrow \langle \psi_{\mathbf{k}_f}^{(-)} | V | \psi_{\mathbf{k}_i}^{(+)} \rangle \qquad ; \text{ distorted-wave B.A.}$$

Here $\psi_{\mathbf{k}}^{(\pm)}(\mathbf{r}_p)$ are the scattering states of the projectile in the nuclear potential with outgoing and incoming-wave boundary conditions.

Solution to Problem 6.8

Instead of computing the transition amplitude with incoming and outgoing plane waves and a single interaction with the potential V (Born approximation), the incoming and outgoing lines are to be dressed with any number of elastic interactions with v on those lines (see Fig. 5.13 in the text). With the use of the rules in Eqs. (5.70)–(5.75), the matrix element of the potential then gets replaced by[8]

$$\langle \mathbf{k}_f | V | \mathbf{k}_i \rangle \to \sum_{n=0}^{\infty} \sum_{m=0}^{\infty} \int d^3x \int d^3x_1 \cdots d^3x_n \int d^3y_1 \cdots d^3y_m$$
$$\times \left[e^{-i\mathbf{k}_f \cdot \mathbf{x}_n} v(x_n) \cdots (-1) G_{k_f}^{(+)}(\mathbf{x}_2 - \mathbf{x}_1) v(x_1)(-1) G_{k_f}^{(+)}(\mathbf{x}_1 - \mathbf{x}) \right] V(\mathbf{x})$$
$$\times \left[(-1) G_{k_i}^{(+)}(\mathbf{x} - \mathbf{y}_1) v(y_1)(-1) G_{k_i}^{(+)}(\mathbf{y}_1 - \mathbf{y}_2) v(y_2) \cdots v(y_m) e^{i\mathbf{k}_i \cdot \mathbf{y}_m} \right]$$

Define the scattering state with outgoing-wave boundary conditions by

$$\psi_{\mathbf{k}_i}^{(+)}(\mathbf{x}) \equiv e^{i\mathbf{k}_i \cdot \mathbf{x}} + \sum_{m=1}^{\infty} \int d^3y_1 \cdots d^3y_m$$
$$\times \left[(-1) G_{k_i}^{(+)}(\mathbf{x} - \mathbf{y}_1) v(y_1)(-1) G_{k_i}^{(+)}(\mathbf{y}_1 - \mathbf{y}_2) v(y_2) \cdots v(y_m) e^{i\mathbf{k}_i \cdot \mathbf{y}_m} \right]$$

This is the iterated form of the integral equation

$$\psi_{\mathbf{k}_i}^{(+)}(\mathbf{x}) = e^{i\mathbf{k}_i \cdot \mathbf{x}} - \int d^3y \, G_{k_i}^{(+)}(\mathbf{x} - \mathbf{y}) v(y) \psi_{\mathbf{k}_i}^{(+)}(\mathbf{y})$$

In a similar fashion, define the scattering state with incoming-wave boundary conditions by

$$\psi_{\mathbf{k}_f}^{(-)}(\mathbf{x}) \equiv e^{i\mathbf{k}_f \cdot \mathbf{x}} + \sum_{n=1}^{\infty} \int d^3x_1 \cdots d^3x_n$$
$$\times \left[(-1) G_{k_f}^{(-)}(\mathbf{x} - \mathbf{x}_1) v(x_1)(-1) G_{k_f}^{(-)}(\mathbf{x}_1 - \mathbf{x}_2) v(x_2) \cdots v(x_n) e^{i\mathbf{k}_f \cdot \mathbf{x}_n} \right]$$

This is the iterated form of the integral equation

$$\psi_{\mathbf{k}_f}^{(-)}(\mathbf{x}) = e^{i\mathbf{k}_f \cdot \mathbf{x}} - \int d^3y \, G_{k_f}^{(-)}(\mathbf{x} - \mathbf{y}) v(y) \psi_{\mathbf{k}_f}^{(-)}(\mathbf{y})$$

Now use the following property of the Green's functions

$$G_k^{(-)}(\mathbf{x}_i - \mathbf{x}_j)^{\star} = G_k^{(+)}(\mathbf{x}_j - \mathbf{x}_i)$$

---

[8]Here, by convention, the $m = 0$ term in the sum is just $e^{i\mathbf{k}_i \cdot \mathbf{x}}$, and the $n = 0$ term is just $e^{-i\mathbf{k}_f \cdot \mathbf{x}}$, where $\mathbf{x}$ now labels the projectile coordinate.

The iterated form of the scattering states then establishes the relation

$$
\langle \psi_{\mathbf{k}_f}^{(-)} | V | \psi_{\mathbf{k}_i}^{(+)} \rangle = \int d^3x \left[\psi_{\mathbf{k}_f}^{(-)}(\mathbf{x}) \right]^* V(\mathbf{x}) \psi_{\mathbf{k}_i}^{(+)}(\mathbf{x})
$$

$$
= \sum_{n=0}^{\infty} \sum_{m=0}^{\infty} \int d^3x \int d^3x_1 \cdots d^3x_n \int d^3y_1 \cdots d^3y_m
$$

$$
\times \left[e^{-i\mathbf{k}_f \cdot \mathbf{x}_n} v(x_n) \cdots (-1) G_{\mathbf{k}_f}^{(+)}(\mathbf{x}_2 - \mathbf{x}_1) v(x_1)(-1) G_{\mathbf{k}_f}^{(+)}(\mathbf{x}_1 - \mathbf{x}) \right] V(\mathbf{x})
$$

$$
\times \left[(-1) G_{\mathbf{k}_i}^{(+)}(\mathbf{x} - \mathbf{y}_1) v(y_1)(-1) G_{\mathbf{k}_i}^{(+)}(\mathbf{y}_1 - \mathbf{y}_2) v(y_2) \cdots v(y_m) e^{i\mathbf{k}_i \cdot \mathbf{y}_m} \right]
$$

Hence the appropriate replacement in distorted-wave Born approximation is

$$
\langle \mathbf{k}_f | V | \mathbf{k}_i \rangle \rightarrow \langle \psi_{\mathbf{k}_f}^{(-)} | V | \psi_{\mathbf{k}_i}^{(+)} \rangle
$$

Problem 6.9 An example with *two* particles in the continuum in the final state is given by charged lepton $l^{\pm}$ scattering from the H-atom, where the atom is ionized. Assume the final atomic electron is in a free plane-wave state with wavenumber $\mathbf{t}$. Thus the initial and final wave functions and energies are

$$
\phi_i = \psi_{1s}(\mathbf{r}_e) \frac{1}{\sqrt{\Omega}} e^{i\mathbf{k}_i \cdot \mathbf{r}_l} \qquad ; \; E_i = \varepsilon_{1s}^0 + \frac{\hbar^2 k_i^2}{2m_l}
$$

$$
\phi_f = \frac{1}{\sqrt{\Omega}} e^{i\mathbf{t} \cdot \mathbf{r}_e} \frac{1}{\sqrt{\Omega}} e^{i\mathbf{k}_f \cdot \mathbf{r}_l} \qquad ; \; E_f = \frac{\hbar^2 t^2}{2m_e} + \frac{\hbar^2 k_f^2}{2m_l}
$$

(a) Show the cross section is given to leading order in H_1 by

$$
d\sigma_{\mathbf{t} \leftarrow 1s} = \frac{2\pi}{\hbar} |\langle \phi_f | H_1 | \phi_i \rangle|^2 \delta(E_f - E_i) \frac{\Omega d^3 t}{(2\pi)^3} \frac{\Omega d^3 k_f}{(2\pi)^3} \frac{1}{I_{\text{inc}}}
$$

(b) Show the quantization volume Ω cancels from this expression;
(c) Do the integral over dt, and show that[9]

$$
d\sigma_{\mathbf{t} \leftarrow 1s} = \frac{4\alpha^2}{q^4} \left(\frac{m_l c}{\hbar} \right) \left(\frac{m_e c}{\hbar} \right) \left(\frac{t}{k_i} \right) |\tilde{\psi}_{1s}(\mathbf{t} + \mathbf{q})|^2 \frac{d^3 k_f}{(2\pi)^3} d\Omega_t
$$

where $\tilde{\psi}_{1s}(\mathbf{t} + \mathbf{q})$ is the Fourier transform of the ground-state wave function

$$
\tilde{\psi}_{1s}(\mathbf{t} + \mathbf{q}) = \int d^3 r_e \, \psi_{1s}(\mathbf{r}_e) e^{-i(\mathbf{t} + \mathbf{q}) \cdot \mathbf{r}_e} \qquad ; \; \mathbf{q} = \mathbf{k}_f - \mathbf{k}_i
$$

---

[9] Now $E_f = E_i$ determines $t(k_f, k_i)$. Recall the caveats in the text if $l^{\pm} = e^{\pm}$.

Solution to Problem 6.9

The solution again closely follows the analysis in section 6.2.2.3. The initial and final states in this case are

$$\phi_i = \psi_{1s}(\mathbf{r}_e)\frac{1}{\sqrt{\Omega}}e^{i\mathbf{k}_i\cdot\mathbf{r}_l} \qquad ; E_i = \varepsilon^0_{1s} + \frac{\hbar^2 k_i^2}{2m_l}$$

$$\phi_f = \frac{1}{\sqrt{\Omega}}e^{it\cdot\mathbf{r}_e}\frac{1}{\sqrt{\Omega}}e^{i\mathbf{k}_f\cdot\mathbf{r}_l} \qquad ; E_f = \frac{\hbar^2 t^2}{2m_e} + \frac{\hbar^2 k_f^2}{2m_l}$$

The effective interaction potential is given in Eq. (6.99)

$$H_1 \doteq -\frac{z_l e^2}{|\mathbf{r}_l - \mathbf{r}_e|}$$

where the charge on the incident lepton is $z_l = \pm 1$.

(a) There are now two particles in the continuum in the final state, and an additional factor $\Omega d^3 t/(2\pi)^3$ is needed in the transition rate to account for those ejected particles that get into the second detector. Thus the cross section becomes

$$d\sigma_{t\leftarrow 1s} = \frac{2\pi}{\hbar}|\langle\phi_f|H_1|\phi_i\rangle|^2\delta(E_f - E_i)\frac{\Omega d^3 t}{(2\pi)^3}\frac{\Omega d^3 k_f}{(2\pi)^3}\frac{1}{I_{inc}}$$

(b) The incident flux is again given by Eq. (6.113), and therefore

$$d\sigma_{t\leftarrow 1s} = \frac{2\pi}{\hbar}|\langle\phi_f|H_1|\phi_i\rangle|^2\delta(E_f - E_i)\frac{\Omega d^3 t}{(2\pi)^3}\frac{\Omega d^3 k_f}{(2\pi)^3}\frac{1}{(\hbar k_i/\Omega m_l)}$$

The matrix element of the interaction follows as in Eqs. (6.102)–(6.103)

$$\langle\phi_f|H_1|\phi_i\rangle = -\frac{4\pi e^2}{q^2}\frac{z_l}{\Omega^{3/2}}\int d^3 r_e\, e^{-i(\mathbf{q}+\mathbf{t})\cdot\mathbf{r}_e}\,\psi_{1s}(\mathbf{r}_e) \qquad ; \mathbf{q} = \mathbf{k}_f - \mathbf{k}_i$$

$$= -\frac{4\pi e^2}{q^2}\frac{z_l}{\Omega^{3/2}}\tilde{\psi}_{1s}(\mathbf{t} + \mathbf{q})$$

The quantization volume Ω now evidently cancels from the cross section, as it must.

(c) The integral over $t^2 dt$ can be carried out making use of the energy-conserving delta-function

$$\int t^2 dt\,\delta(E_f - E_i) = \frac{m_e}{\hbar^2}\int t\,d\left(\frac{\hbar^2 t^2}{2m_e}\right)\delta(E_f - E_i) = \frac{m_e t}{\hbar^2} \qquad ; E_f = E_i$$

A combination of these results gives

$$d\sigma_{t\leftarrow 1s} = \frac{4\alpha^2}{q^4}\left(\frac{m_l c}{\hbar}\right)\left(\frac{m_e c}{\hbar}\right)\left(\frac{t}{k_i}\right)|\tilde{\psi}_{1s}(\mathbf{t}+\mathbf{q})|^2 \frac{d^3 k_f}{(2\pi)^3}d\Omega_t$$

where $\alpha = e^2/\hbar c$ is the fine-structure constant. Note that this experiment now measures the Fourier transform of the ground-state wave function.

Problem 6.10 A typical nuclear gamma transition is of the order of 1 MeV. The nuclear radius is $R \approx 1.2B^{1/3} \times 10^{-13}$ cm, where B is the baryon number.
(a) Compare the nuclear size with the wavelength of the radiation;
(b) Compare with the atomic case.

Solution to Problem 6.10

(a) Use

$$\hbar c = 197.3 \text{ MeV-F} = 197.3 \times 10^{-13} \text{ MeV-cm}$$

For a nuclear transition with $\Delta E = h\nu = 1$ MeV, the wavelength is

$$\lambda = \frac{c}{\nu} = 2\pi\frac{\hbar c}{\Delta E} = 1.240 \times 10^{-10} \text{ cm}$$

The nuclear radius is $R \approx 1.2B^{1/3} \times 10^{-13}$ cm. Thus in the nuclear case

$$\frac{R}{\lambda} \approx 0.97B^{1/3} \times 10^{-3}$$

Even for $^{208}\text{Pb}_{82}$, this number is $R/\lambda \approx 5.75 \times 10^{-3}$.
(b) Take $\Delta E = 1$ eV in the atomic case. Then

$$\lambda = \frac{c}{\nu} = 2\pi\frac{\hbar c}{\Delta E} = 1.240 \times 10^{-4} \text{ cm}$$

Assume the atomic radius is characterized by $R \approx a_0/Z = 0.5292\,Z^{-1} \times 10^{-8}$ cm. Thus in the atomic case

$$\frac{R}{\lambda} \approx 4.27Z^{-1} \times 10^{-5}$$

For hydrogen, this number is $R/\lambda \approx 4.27 \times 10^{-5}$.

Problem 6.11 Start from Eqs. (6.130), and define the probability density and probability current by[10]

$$\rho \equiv \Psi^* \Psi$$

$$\mathbf{S} \equiv \frac{1}{2m} \left\{ \Psi^* \left(\mathbf{p} - \frac{e}{c} \mathbf{A} \right) \Psi + \left[\left(\mathbf{p} - \frac{e}{c} \mathbf{A} \right) \Psi \right]^* \Psi \right\}$$

Work in the coordinate representation, and show this current is conserved

$$\boldsymbol{\nabla} \cdot \mathbf{S} + \frac{\partial \rho}{\partial t} = 0$$

As before, these equations provide the interpretation for a particle in an external electromagnetic field.

Solution to Problem 6.11

We use the Schrödinger equation together with its complex conjugate

$$i\hbar \frac{\partial \Psi}{\partial t} = \left[\frac{1}{2m} \left(\mathbf{p} - \frac{e}{c} \mathbf{A} \right)^2 + e\Phi + V \right] \Psi$$

$$-i\hbar \frac{\partial \Psi^*}{\partial t} = \left\{ \left[\frac{1}{2m} \left(\mathbf{p} - \frac{e}{c} \mathbf{A} \right)^2 + e\Phi + V \right] \Psi \right\}^*$$

If we multiply the first equation by Ψ^*, the second equation by Ψ, and take the difference, we obtain

$$i\hbar \frac{\partial}{\partial t} |\Psi|^2 = \frac{1}{2m} \left\{ \Psi^* \left(\mathbf{p} - \frac{e}{c} \mathbf{A} \right)^2 \Psi - \left[\left(\mathbf{p} - \frac{e}{c} \mathbf{A} \right)^2 \Psi \right]^* \Psi \right\}$$

We now need to check that the r.h.s. can be written as a divergence. We have

$$\text{r.h.s.} = \frac{1}{2m} \left\{ \Psi^* \left(\mathbf{p} - \frac{e}{c} \mathbf{A} \right)^2 \Psi - \left[\left(\mathbf{p} - \frac{e}{c} \mathbf{A} \right)^2 \Psi \right]^* \Psi \right\}$$

$$= \frac{\hbar}{2mi} \left\{ \Psi^* \boldsymbol{\nabla} \cdot \left(\mathbf{p} - \frac{e}{c} \mathbf{A} \right) \Psi + \left[\boldsymbol{\nabla} \cdot \left(\mathbf{p} - \frac{e}{c} \mathbf{A} \right) \Psi \right]^* \Psi \right\}$$

$$\quad - \frac{e}{2mc} \left\{ \Psi^* \mathbf{A} \cdot \left(\mathbf{p} - \frac{e}{c} \mathbf{A} \right) \Psi - \left[\mathbf{A} \cdot \left(\mathbf{p} - \frac{e}{c} \mathbf{A} \right) \Psi \right]^* \Psi \right\}$$

$$\equiv (A) + (B)$$

---

[10]Note that $(\mathbf{p} - e\mathbf{A}/c)/m = \mathbf{v}$ is the particle's *kinetic velocity* (see appendix B). Recall that here we work in c.g.s. units.

Let us first separately simplify the two parts of r.h.s.

$$(A) = \frac{\hbar}{2mi}\left\{\Psi^*\boldsymbol{\nabla}\cdot\left(\mathbf{p}-\frac{e}{c}\mathbf{A}\right)\Psi + \left[\boldsymbol{\nabla}\cdot\left(\mathbf{p}-\frac{e}{c}\mathbf{A}\right)\Psi\right]^*\Psi\right\}$$

$$= \frac{\hbar}{2mi}\boldsymbol{\nabla}\cdot\left\{\Psi^*\left(\mathbf{p}-\frac{e}{c}\mathbf{A}\right)\Psi + \left[\left(\mathbf{p}-\frac{e}{c}\mathbf{A}\right)\Psi\right]^*\Psi\right\}$$

$$-\frac{\hbar}{2mi}\left\{\boldsymbol{\nabla}\Psi^*\cdot\left(\mathbf{p}-\frac{e}{c}\mathbf{A}\right)\Psi + \left[\left(\mathbf{p}-\frac{e}{c}\mathbf{A}\right)\Psi\right]^*\cdot\boldsymbol{\nabla}\Psi\right\}$$

$$= \frac{\hbar}{2mi}\boldsymbol{\nabla}\cdot\left\{\Psi^*\left(\mathbf{p}-\frac{e}{c}\mathbf{A}\right)\Psi + \left[\left(\mathbf{p}-\frac{e}{c}\mathbf{A}\right)\Psi\right]^*\Psi\right\}$$

$$+\frac{e\hbar}{2mci}\left(\boldsymbol{\nabla}\Psi^*\cdot\mathbf{A}\Psi + \Psi^*\mathbf{A}\cdot\boldsymbol{\nabla}\Psi\right)$$

Furthermore

$$(B) = -\frac{e}{2mc}\left\{\Psi^*\mathbf{A}\cdot\left(\mathbf{p}-\frac{e}{c}\mathbf{A}\right)\Psi - \left[\mathbf{A}\cdot\left(\mathbf{p}-\frac{e}{c}\mathbf{A}\right)\Psi\right]^*\Psi\right\}$$

$$= -\frac{e\hbar}{2mci}\left[\Psi^*\mathbf{A}\cdot\boldsymbol{\nabla}\Psi + \boldsymbol{\nabla}\Psi^*\cdot\mathbf{A}\Psi\right]$$

As a result we have

$$\text{r.h.s.} = (A) + (B)$$

$$= \frac{\hbar}{2mi}\boldsymbol{\nabla}\cdot\left\{\Psi^*\left(\mathbf{p}-\frac{e}{c}\mathbf{A}\right)\Psi + \left[\left(\mathbf{p}-\frac{e}{c}\mathbf{A}\right)\Psi\right]^*\Psi\right\}$$

Thus we obtain the continuity equation for a particle in an electromagnetic field

$$\frac{\partial\rho}{\partial t} + \boldsymbol{\nabla}\cdot\mathbf{S} = 0$$

where the density and current are given by

$$\rho = |\Psi|^2$$

$$\mathbf{S} = \frac{1}{2m}\left\{\Psi^*\left(\mathbf{p}-\frac{e}{c}\mathbf{A}\right)\Psi + \left[\left(\mathbf{p}-\frac{e}{c}\mathbf{A}\right)\Psi\right]^*\Psi\right\}$$

Problem 6.12 Make a unitary transformation from the Schrödinger to the interaction picture

$$|\Psi_I(t)\rangle = e^{i\hat{H}_0 t/\hbar}|\Psi(t)\rangle \qquad ; \text{ interaction picture}$$

$$\hat{H}_I(t) = e^{i\hat{H}_0 t/\hbar}\hat{H}_1(t)e^{-i\hat{H}_0 t/\hbar}$$

(a) Show the Schrödinger equation in the interaction picture is

$$i\hbar\frac{\partial}{\partial t}|\Psi_I(t)\rangle = \hat{H}_I(t)|\Psi_I(t)\rangle \qquad ; \text{S-eqn}$$

(b) Show the time-development operator $\hat{U}(t, t_0)$ in Eq. (6.189) satisfies

$$i\hbar\frac{\partial}{\partial t}\hat{U}(t, t_0) = \hat{H}_I(t)\hat{U}(t, t_0)$$

$$\hat{U}(t_0, t_0) = 1$$

(c) Hence show that $|\Psi_I(t)\rangle = \hat{U}(t, t_0)|\Psi_I(t_0)\rangle$ provides the solution to the Schrödinger equation in the interaction picture that reduces to $|\Psi_I(t_0)\rangle$ at $t = t_0$.

Solution to Problem 6.12

(a) Substitute the above in the Schrödinger equation

$$i\hbar\frac{\partial}{\partial t}|\Psi(t)\rangle = \hat{H}|\Psi(t)\rangle$$

This leads to

$$i\hbar\frac{\partial}{\partial t}\left[e^{-i\hat{H}_0 t/\hbar}|\Psi_I(t)\rangle\right] = \hat{H}_0 e^{-i\hat{H}_0 t/\hbar}|\Psi_I(t)\rangle + e^{-i\hat{H}_0 t/\hbar}i\hbar\frac{\partial}{\partial t}|\Psi_I(t)\rangle$$

$$= (\hat{H}_0 + \hat{H}_1)e^{-i\hat{H}_0 t/\hbar}|\Psi_I(t)\rangle$$

A cancellation of terms, and multiplication be $e^{i\hat{H}_0 t/\hbar}$ then gives

$$i\hbar\frac{\partial}{\partial t}|\Psi_I(t)\rangle = \hat{H}_I(t)|\Psi_I(t)\rangle$$

$$\hat{H}_I(t) = e^{i\hat{H}_0 t/\hbar}\hat{H}_1(t)e^{-i\hat{H}_0 t/\hbar}$$

(b) Equation (6.189) expresses the time-development operator $\hat{\mathcal{U}}(t, t_0)$ as a series[11]

$$\hat{\mathcal{U}}(t, t_0) = \sum_{n=0}^{\infty}\left(-\frac{i}{\hbar}\right)^n \int_{t_0}^t dt_1 \int_{t_0}^{t_1} dt_2 \cdots \int_{t_0}^{t_{n-1}} dt_n$$

$$\times \hat{H}_I(t_1)\hat{H}_I(t_2)\cdots\hat{H}_I(t_n)$$

---

[11]The $n = 0$ term in the series is 1.

The time derivative of this expression is

$$i\hbar\frac{\partial}{\partial t}\hat{U}(t,t_0) = \hat{H}_I(t) \sum_{n=0}^{\infty} \left(-\frac{i}{\hbar}\right)^n \int_{t_0}^{t} dt_1 \int_{t_0}^{t_1} dt_2 \cdots \int_{t_0}^{t_{n-1}} dt_n$$
$$\times \hat{H}_I(t_1)\hat{H}_I(t_2)\cdots\hat{H}_I(t_n)$$

Hence

$$i\hbar\frac{\partial}{\partial t}\hat{U}(t,t_0) = \hat{H}_I(t)\hat{U}(t,t_0)$$

Furthermore, it is evident from the series that

$$\hat{U}(t_0,t_0) = 1$$

(c) Thus the following

$$|\Psi_I(t)\rangle = \hat{U}(t,t_0)|\Psi_I(t_0)\rangle$$

provides the solution to the Schrödinger equation in the interaction picture that reduces to $|\Psi_I(t_0)\rangle$ at $t = t_0$.

Chapter 7

Electromagnetic Radiation and Quantum Electrodynamics

Problem 7.1 Prove that the expansion of the vector potential $\mathbf{A}(\mathbf{x}, t)$ in Eq. (7.23) has enough flexibility to match the arbitrary set of initial conditions in Eqs. (7.26).[1]

Solution to Problem 7.1

The expansion of the vector potential in Eq. (7.23) is

$$\mathbf{A}(\mathbf{x}, t) = \frac{1}{\sqrt{\Omega}} \sum_{\mathbf{k}} \sum_{s=1}^{2} \left[a(\mathbf{k}, s) \mathbf{e}_{\mathbf{k},s} \, e^{i(\mathbf{k} \cdot \mathbf{x} - \omega_k t)} + a(\mathbf{k}, s)^{\star} \mathbf{e}_{\mathbf{k},s} \, e^{-i(\mathbf{k} \cdot \mathbf{x} - \omega_k t)} \right]$$

At an initial time $t = 0$, this quantity and its time derivative are given by

$$\mathbf{A}(\mathbf{x}, 0) = \frac{1}{\sqrt{\Omega}} \sum_{\mathbf{k}} \sum_{s=1}^{2} \left[a(\mathbf{k}, s) \mathbf{e}_{\mathbf{k},s} \, e^{i\mathbf{k} \cdot \mathbf{x}} + a(\mathbf{k}, s)^{\star} \mathbf{e}_{\mathbf{k},s} \, e^{-i\mathbf{k} \cdot \mathbf{x}} \right]$$

$$\frac{\partial \mathbf{A}(\mathbf{x}, 0)}{\partial t} = \frac{1}{\sqrt{\Omega}} \sum_{\mathbf{k}} \sum_{s=1}^{2} (-i\omega_k) \left[a(\mathbf{k}, s) \mathbf{e}_{\mathbf{k},s} \, e^{i\mathbf{k} \cdot \mathbf{x}} - a(\mathbf{k}, s)^{\star} \mathbf{e}_{\mathbf{k},s} \, e^{-i\mathbf{k} \cdot \mathbf{x}} \right]$$

Take the inverse Fourier transform of these expressions using Eq. (7.30), divide the second by $(-i\omega_k)$, and dot the results into $\mathbf{e}_{\mathbf{k},s}$

$$\frac{1}{\sqrt{\Omega}} \int d^3x \, e^{-i\mathbf{k} \cdot \mathbf{x}} \, \mathbf{e}_{\mathbf{k},s} \cdot \mathbf{A}(\mathbf{x}, 0) = a(\mathbf{k}, s) + \sum_{s'} a(-\mathbf{k}, s')^{\star} (\mathbf{e}_{-\mathbf{k},s'} \cdot \mathbf{e}_{\mathbf{k},s})$$

$$\frac{1}{\sqrt{\Omega}} \int d^3x \, e^{-i\mathbf{k} \cdot \mathbf{x}} \, \mathbf{e}_{\mathbf{k},s} \cdot \left(\frac{i}{\omega_k} \right) \frac{\partial \mathbf{A}(\mathbf{x}, 0)}{\partial t} = a(\mathbf{k}, s) - \sum_{s'} a(-\mathbf{k}, s')^{\star} (\mathbf{e}_{-\mathbf{k},s'} \cdot \mathbf{e}_{\mathbf{k},s})$$

---

[1] Compare Prob. 4.2 in Vol. I.

Now take 1/2 the sum to solve for the Fourier coefficient $a(\mathbf{k}, s)$

$$a(\mathbf{k}, s) = \frac{1}{\sqrt{\Omega}} \int d^3x \, e^{-i\mathbf{k}\cdot\mathbf{x}} \, \mathbf{e}_{\mathbf{k},s} \cdot \frac{1}{2} \left[\mathbf{A}(\mathbf{x}, 0) + \left(\frac{i}{\omega_k} \right) \frac{\partial \mathbf{A}(\mathbf{x}, 0)}{\partial t} \right]$$

When inserted back in the field expansion, this determines $\mathbf{A}(\mathbf{x}, t)$ for all subsequent time in terms of the initial values of the field and its time derivative.

Problem 7.2 Start from Eq. (7.57). Introduce the expansion of $\mathbf{A}(\mathbf{x}, t)$ in Eq. (7.36), and derive the expression for the momentum in the normal modes of the free, classical radiation field in Eq. (7.60).[2]

Solution to Problem 7.2

We start from Eq. (7.57)

$$\mathbf{P} = -\frac{1}{4\pi c^2} \int_\Omega d^3x \left(\frac{\partial \mathbf{A}}{\partial t} \right) \times (\mathbf{\nabla} \times \mathbf{A})$$

where the vector potential is given by Eq. (7.36)

$$\mathbf{A}(\mathbf{x}, t) = \frac{1}{\sqrt{\Omega}} \sum_{\mathbf{k}} \sum_{s=1}^{2} \left(\frac{2\pi\hbar c^2}{\omega_k} \right)^{1/2} \mathbf{e}_{\mathbf{k}s} \left(a_{\mathbf{k}s} \, e^{i\mathbf{k}\cdot\mathbf{x}-i\omega_k t} + a_{\mathbf{k}s}^{\star} \, e^{-i\mathbf{k}\cdot\mathbf{x}+i\omega_k t} \right)$$

Insertion in the above gives

$$\mathbf{P} = -\frac{2\pi\hbar c^2}{4\pi c^2 \Omega} \int_\Omega d^3x \sum_{\mathbf{k}} \sum_{\mathbf{k}'} \sum_{s,s'} \left(\frac{\omega_k}{\omega_{k'}} \right)^{1/2} [\mathbf{e}_{\mathbf{k}s} \times (\mathbf{k}' \times \mathbf{e}_{\mathbf{k}'s'})]$$

$$\times \left(a_{\mathbf{k}s} \, e^{i\mathbf{k}\cdot\mathbf{x}-i\omega_k t} - a_{\mathbf{k}s}^{\star} \, e^{-i\mathbf{k}\cdot\mathbf{x}+i\omega_k t} \right) \left(a_{\mathbf{k}'s'} \, e^{i\mathbf{k}'\cdot\mathbf{x}-i\omega_{k'} t} - a_{\mathbf{k}'s'}^{\star} \, e^{-i\mathbf{k}'\cdot\mathbf{x}+i\omega_{k'} t} \right)$$

Integration over the volume Ω with periodic boundary conditions then gives

$$\mathbf{P} = \frac{\hbar}{2} \sum_{\mathbf{k}} \sum_{s,s'} \{ \mathbf{k} \, \delta_{ss'} (a_{\mathbf{k}s} a_{\mathbf{k}s}^{\star} + a_{\mathbf{k}s}^{\star} a_{\mathbf{k}s})$$

$$+ \mathbf{k}(\mathbf{e}_{\mathbf{k}s} \cdot \mathbf{e}_{-\mathbf{k},s'}) \left(a_{\mathbf{k}s} a_{-\mathbf{k},s'} \, e^{-2i\omega_k t} + a_{\mathbf{k}s}^{\star} a_{-\mathbf{k},s'}^{\star} \, e^{2i\omega_k t} \right) \}$$

Here use has been made of Eq. (7.58)

$$\mathbf{e}_{\mathbf{k}s} \times (\mathbf{k} \times \mathbf{e}_{\mathbf{k}s'}) = \mathbf{k} \, (\mathbf{e}_{\mathbf{k}s} \cdot \mathbf{e}_{\mathbf{k}s'}) = \mathbf{k} \, \delta_{s,s'}$$

$$\mathbf{e}_{\mathbf{k}s} \times (\mathbf{k} \times \mathbf{e}_{-\mathbf{k},s'}) = \mathbf{k} \, (\mathbf{e}_{\mathbf{k}s} \cdot \mathbf{e}_{-\mathbf{k},s'})$$

---

[2] Recall Eq. (7.34).

It is observed in Eq. (7.59) that the last term in **P** vanishes, leading to

$$\mathbf{P} = \frac{1}{2} \sum_{\mathbf{k}} \sum_{s=1}^{2} \hbar \mathbf{k} \left(a_{\mathbf{k}s} a_{\mathbf{k}s}^* + a_{\mathbf{k}s}^* a_{\mathbf{k}s} \right)$$

which is just Eq. (7.60).

Problem 7.3 (a) Establish the vector identity

$$\nabla \cdot (\mathbf{a} \times \mathbf{b}) = \mathbf{b} \cdot (\nabla \times \mathbf{a}) - \mathbf{a} \cdot (\nabla \times \mathbf{b})$$

(b) Integrate the Poynting vector in Eq. (7.55) over the surface of a small volume in space. Use Gauss's law and Maxwell's equations to reproduce the energy density in the free radiation field in Eq. (7.27).

Solution to Problem 7.3

(a) We first prove the vector identity. Write

$$\nabla \cdot (\mathbf{a} \times \mathbf{b}) = \sum_{i=1}^{3} \sum_{j=1}^{3} \sum_{k=1}^{3} \varepsilon_{ijk} \frac{\partial}{\partial x_i} a_j b_k$$

where ε_{ijk} is the completely antisymmetric Levi-Civita tensor. Then

$$\nabla \cdot (\mathbf{a} \times \mathbf{b}) = \sum_i \sum_j \sum_k \varepsilon_{ijk} \left[\frac{\partial a_j}{\partial x_i} b_k + a_j \frac{\partial b_k}{\partial x_i} \right]$$

Hence

$$\nabla \cdot (\mathbf{a} \times \mathbf{b}) = \mathbf{b} \cdot (\nabla \times \mathbf{a}) - \mathbf{a} \cdot (\nabla \times \mathbf{b})$$

(b) The Poynting vector is given in Eq. (7.55)

$$\mathbf{S}_\gamma = \frac{c}{4\pi} \boldsymbol{\mathcal{E}} \times \mathbf{B}$$

This is the *energy flux* in the radiation field. From the above identity

$$\nabla \cdot \mathbf{S}_\gamma = \frac{c}{4\pi} \left[\mathbf{B} \cdot (\nabla \times \boldsymbol{\mathcal{E}}) - \boldsymbol{\mathcal{E}} \cdot (\nabla \times \mathbf{B}) \right]$$

Maxwell's Eqs. (7.1) in free space with no sources imply

$$\nabla \times \boldsymbol{\mathcal{E}} = -\frac{1}{c} \frac{\partial \mathbf{B}}{\partial t}$$

$$\nabla \times \mathbf{B} = \frac{1}{c} \frac{\partial \boldsymbol{\mathcal{E}}}{\partial t}$$

Thus

$$\nabla \cdot \mathbf{S}_\gamma = -\frac{1}{4\pi}\left[\mathbf{B}\cdot\frac{\partial \mathbf{B}}{\partial t} + \boldsymbol{\mathcal{E}}\cdot\frac{\partial \boldsymbol{\mathcal{E}}}{\partial t}\right]$$

$$= -\frac{\partial}{\partial t}\left[\frac{1}{8\pi}\left(\boldsymbol{\mathcal{E}}^2 + \mathbf{B}^2\right)\right]$$

Now compute the net outflow of the Poynting vector from a small sphere, and use Gauss's theorem to re-write it as a volume integral

$$\int_A d\mathbf{A}\cdot\mathbf{S}_\gamma = \int_V d^3x\,\nabla\cdot\mathbf{S}_\gamma$$

$$= -\frac{d}{dt}\int_V d^3x\left[\frac{1}{8\pi}\left(\boldsymbol{\mathcal{E}}^2 + \mathbf{B}^2\right)\right]$$

We interpret this as the rate of decrease of the energy in the volume, where the *energy* in the radiation field is given by Eq. (7.27).

Problem 7.4 (a) Prove the relation in Eq. (7.66), which states that at equal times the commutator of the vector field operator at two distinct spatial points vanishes;

(b) Use this to conclude that starting from Eq. (7.73), one has

$$\dot{\hat{\mathbf{A}}}(\mathbf{x}) = \frac{i}{\hbar}[\hat{H}, \hat{\mathbf{A}}(\mathbf{x})] = \frac{i}{\hbar}[\hat{H}_{\rm rad}, \hat{\mathbf{A}}(\mathbf{x})]$$

Solution to Problem 7.4

(a) From Eq. (7.45), the vector field operator in the Coulomb gauge in the Schrödinger picture is

$$\hat{\mathbf{A}}(\mathbf{x}) \equiv \frac{1}{\sqrt{\Omega}}\sum_{\mathbf{k}}\sum_{s=1}^{2}\left(\frac{2\pi\hbar c^2}{\omega_k}\right)^{1/2}\mathbf{e}_{\mathbf{k}s}\left(\hat{a}_{\mathbf{k}s}\,e^{i\mathbf{k}\cdot\mathbf{x}} + \hat{a}^\dagger_{\mathbf{k}s}\,e^{-i\mathbf{k}\cdot\mathbf{x}}\right)$$

where the creation and destruction operators satisfy the canonical commutation relations in Eqs. (7.38). The commutator of the field at two distinct spatial points is then given by

$$[\hat{A}_i(\mathbf{x}),\,\hat{A}_j(\mathbf{x}')] = \sum_{\mathbf{k},s}\left(\frac{2\pi\hbar c^2}{\Omega\omega_k}\right)(\mathbf{e}_{\mathbf{k}s})_i(\mathbf{e}_{\mathbf{k}s})_j\left[e^{i\mathbf{k}\cdot(\mathbf{x}-\mathbf{x}')} - e^{-i\mathbf{k}\cdot(\mathbf{x}-\mathbf{x}')}\right]$$

The sum over polarizations is done in Eqs. (7.64)

$$\sum_{s=1}^{2}(\mathbf{e}_{\mathbf{k}s})_i(\mathbf{e}_{\mathbf{k}s})_j = \delta_{ij} - \frac{k_ik_j}{\mathbf{k}^2}$$

Thus

$$[\hat{A}_i(\mathbf{x}), \hat{A}_j(\mathbf{x}')] = \sum_{\mathbf{k}} \left(\frac{2\pi\hbar c^2}{\Omega\omega_k}\right)\left(\delta_{ij} - \frac{k_ik_j}{\mathbf{k}^2}\right)\left[e^{i\mathbf{k}\cdot(\mathbf{x}-\mathbf{x}')} - e^{-i\mathbf{k}\cdot(\mathbf{x}-\mathbf{x}')}\right]$$

Now change dummy summation variable $\mathbf{k} \to -\mathbf{k}$ for the sum over the second exponential. The two terms then cancel, establishing the result that

$$[\hat{A}_i(\mathbf{x}), \hat{A}_j(\mathbf{x}')] = 0$$

(b) If one computes the commutator of the hamiltonian in Eq. (7.73) with the field $\hat{\mathbf{A}}(\mathbf{x})$ at the spatial point $\mathbf{x}$, making use of the result in part (a), the only term that fails to commute with this field is $\hat{H}_{\mathrm{rad}}$.[3] Therefore

$$\dot{\hat{\mathbf{A}}}(\mathbf{x}) = \frac{i}{\hbar}[\hat{H}, \hat{\mathbf{A}}(\mathbf{x})] = \frac{i}{\hbar}[\hat{H}_{\mathrm{rad}}, \hat{\mathbf{A}}(\mathbf{x})]$$

Problem 7.5 (a) Take the matrix element of the electromagnetic current operator $\hat{\mathbf{j}}(\mathbf{x})$ in Eq. (7.96) between the states $|\phi_m; n_1 \cdots, n_\infty\rangle$. Use the hermiticity of $\hat{\mathbf{x}}_e$, and show the result is

$$\langle\phi_m; n_1 \cdots n_\infty|\hat{\mathbf{j}}(\mathbf{x})|\phi_m; n_1 \cdots n_\infty\rangle = \frac{e}{2c}\{\phi_m^*(\mathbf{x})\dot{\mathbf{x}}\,\phi_m(\mathbf{x}) + [\dot{\mathbf{x}}\,\phi_m(\mathbf{x})]^*\phi_m(\mathbf{x})\}$$

Define all quantities;

(b) Compare with the results in Prob. 6.11. Show the charge density and electromagnetic current are related to the probability density and probability current by

$$\varrho(\mathbf{x}) = e\rho(\mathbf{x}) \qquad ; \hat{\mathbf{j}}(\mathbf{x}) = \frac{e}{c}\hat{\mathbf{S}}(\mathbf{x})$$

Solution to Problem 7.5

(a) The electromagnetic current operator in Eq. (7.96) is given by

$$\hat{\mathbf{j}}(\mathbf{x}) = \frac{e}{c}\left\{\hat{\mathbf{x}}_e, \delta^{(3)}(\mathbf{x} - \mathbf{x}_e)\right\}_{\mathrm{sym}}$$

---

[3]Note that $\mathbf{p}_e = (\hbar/i)\nabla_e$.

where the kinetic velocity operator is given in Eq. (7.76) as

$$\hat{\dot{\mathbf{x}}}_e = \frac{1}{m_0}\left[\mathbf{p}_e - \frac{e}{c}\hat{\mathbf{A}}(\mathbf{x}_e)\right]$$

In the coordinate representation for the particle, where $\mathbf{p}_e = (\hbar/i)\nabla_e$, the matrix element of the current between the states $|\phi_m; n_1 \cdots, n_\infty\rangle$ is given by

$$\langle\phi_m; n_1 \cdots n_\infty|\hat{\mathbf{j}}(\mathbf{x})|\phi_m; n_1 \cdots n_\infty\rangle = \frac{e}{2c}\int d^3x_e$$

$$\times \langle n_1 \cdots n_\infty|\phi_m^\star(\mathbf{x}_e)\left\{\hat{\dot{\mathbf{x}}}_e\delta^{(3)}(\mathbf{x}-\mathbf{x}_e) + \delta^{(3)}(\mathbf{x}-\mathbf{x}_e)\hat{\dot{\mathbf{x}}}_e\right\}\phi_m(\mathbf{x}_e)|n_1 \cdots, n_\infty\rangle$$

The field operator $\hat{\mathbf{A}}(\mathbf{x}_e)$ is given in Eq. (7.68). Since it changes the photon number by one, it has no diagonal matrix element between the states $|n_1 \cdots, n_\infty\rangle$. Hence, in this expression, $\hat{\dot{\mathbf{x}}}_e \doteq \mathbf{p}_e/m_0 \equiv \dot{\mathbf{x}}_e$.

Use the hermiticity of $\dot{\mathbf{x}}_e$ to re-write the above expression as

$$\langle\phi_m; n_1 \cdots n_\infty|\hat{\mathbf{j}}(\mathbf{x})|\phi_m; n_1 \cdots n_\infty\rangle = \frac{e}{2c}\int d^3x_e$$

$$\times \left\{\phi_m^\star(\mathbf{x}_e)\delta^{(3)}(\mathbf{x}-\mathbf{x}_e)[\dot{\mathbf{x}}_e\phi_m(\mathbf{x}_e)] + [\dot{\mathbf{x}}_e\phi_m(\mathbf{x}_e)]^\star \delta^{(3)}(\mathbf{x}-\mathbf{x}_e)\phi_m(\mathbf{x}_e)\right\}$$

where the order of the two terms has been interchanged. The integral over $\int d^3x_e$ can now be carried out to give

$$\langle\phi_m; n_1 \cdots n_\infty|\hat{\mathbf{j}}(\mathbf{x})|\phi_m; n_1 \cdots n_\infty\rangle = \frac{e}{2c}\left\{\phi_m^\star(\mathbf{x})\dot{\mathbf{x}}\,\phi_m(\mathbf{x}) + [\dot{\mathbf{x}}\,\phi_m(\mathbf{x})]^\star\phi_m(\mathbf{x})\right\}$$

which is the stated answer.

(b) A comparison of the above with the results in Prob. 6.11 shows, at least in this example, that the charge density and electromagnetic current are related to the probability density and probability current by

$$\varrho(\mathbf{x}) = e\rho(\mathbf{x}) \qquad\qquad ; \hat{\mathbf{j}}(\mathbf{x}) = \frac{e}{c}\hat{\mathbf{S}}(\mathbf{x})$$

Problem 7.6 Show the transverse delta-function in Eq. (7.65) projects from a vector field $\mathbf{v}(\mathbf{x})$ that part that is *transverse*[4]

$$\mathbf{v}_T(\mathbf{x}) \equiv \int d^3x'\, \underline{\underline{\boldsymbol{\delta}}}^T(\mathbf{x}-\mathbf{x}')\cdot\mathbf{v}(\mathbf{x}')$$

$$\nabla\cdot\mathbf{v}_T(\mathbf{x}) = 0$$

---

[4]The dyadic product is defined by $[\underline{\mathbf{a}}\cdot\mathbf{b}]_i \equiv a_{ij}b_j$. (Recall that repeated Latin indices are summed from 1 to 3.)

Solution to Problem 7.6

An arbitrary vector field has a Fourier series representation

$$\mathbf{v}(\mathbf{x}) = \frac{1}{\sqrt{\Omega}} \sum_{\mathbf{k}} \mathbf{v}(\mathbf{k}) \, e^{i\mathbf{k}\cdot\mathbf{x}}$$

The vector coefficients $\mathbf{v}(\mathbf{k})$ can be expanded in a complete, orthonormal set of unit vectors $\mathbf{e}_{\mathbf{k}s}$, where $\mathbf{e}_{\mathbf{k}1}$ and $\mathbf{e}_{\mathbf{k}2}$ are two unit vectors orthogonal to $\mathbf{k}$ (see Fig. 7.1 in the text), and $\mathbf{e}_{\mathbf{k}0} = \mathbf{k}/k$

$$\mathbf{v}(\mathbf{k}) = \sum_{s=0}^{2} v(\mathbf{k}, s) \, \mathbf{e}_{\mathbf{k}s}$$

The vector field is then separated, quite generally, into transverse and longitudinal parts

$$\mathbf{v}(\mathbf{x}) = \mathbf{v}_T(\mathbf{x}) + \mathbf{v}_L(\mathbf{x})$$

where

$$\mathbf{v}_T(\mathbf{x}) = \frac{1}{\sqrt{\Omega}} \sum_{\mathbf{k}} \sum_{s=1}^{2} v(\mathbf{k}, s) \, \mathbf{e}_{\mathbf{k}s} \, e^{i\mathbf{k}\cdot\mathbf{x}} \qquad ; \; \boldsymbol{\nabla} \cdot \mathbf{v}_T(\mathbf{x}) = 0$$

$$\mathbf{v}_L(\mathbf{x}) = \frac{1}{\sqrt{\Omega}} \sum_{\mathbf{k}} v(\mathbf{k}, 0) \, \mathbf{e}_{\mathbf{k}0} \, e^{i\mathbf{k}\cdot\mathbf{x}} \qquad ; \; \boldsymbol{\nabla} \times \mathbf{v}_L(\mathbf{x}) = 0$$

The transverse delta-function is given in Eqs. (7.65)

$$\delta_{ij}^{T}(\mathbf{x} - \mathbf{x}') = \frac{1}{\Omega} \sum_{\mathbf{k}} \left(\delta_{ij} - \frac{k_i k_j}{k^2} \right) e^{i\mathbf{k}\cdot(\mathbf{x}-\mathbf{x}')}$$

Substitute these expressions into the integral, and use the relation

$$\frac{1}{\Omega} \int d^3x' \, e^{i(\mathbf{k}-\mathbf{k}')\cdot\mathbf{x}'} = \delta_{\mathbf{k},\mathbf{k}'}$$

This gives

$$\int d^3x' \, \delta_{ij}^{T}(\mathbf{x} - \mathbf{x}') v_j(\mathbf{x}') = \frac{1}{\sqrt{\Omega}} \sum_{\mathbf{k}} \sum_{s=0}^{2} v(\mathbf{k}, s) \left(\delta_{ij} - \frac{k_i k_j}{k^2} \right) (\mathbf{e}_{\mathbf{k}s})_j \, e^{i\mathbf{k}\cdot\mathbf{x}}$$

Now use

$$\left(\delta_{ij} - \frac{k_i k_j}{k^2} \right) (\mathbf{e}_{\mathbf{k}s})_j = (\mathbf{e}_{\mathbf{k}s})_i \qquad ; \; s = 1, 2$$

$$= 0 \qquad ; \; s = 0$$

Therefore

$$\int d^3x' \, \underline{\underline{\delta}}^T (\mathbf{x} - \mathbf{x}') \cdot \mathbf{v}(\mathbf{x}') = \frac{1}{\sqrt{\Omega}} \sum_{\mathbf{k}} \sum_{s=1}^{2} v(\mathbf{k}, s) \mathbf{e}_{\mathbf{k}s} \, e^{i\mathbf{k}\cdot\mathbf{x}}$$

$$= \mathbf{v}_T(\mathbf{x})$$

Problem 7.7 (a) Show that the classical field vanishes for a state with a definite number of photons;

(b) Show that there is an uncertainty relation between field strength and photon number [recall Eq. (3.220)];

(c) Show that with many photons in the mode $\{\mathbf{k}, s\}$, produced, for example, by a laser, one can replace the creation and destruction operators for that mode by c-numbers[5]

$$\left(\frac{2\pi\hbar c^2}{\omega_k}\right)^{1/2} \hat{a}_{\mathbf{k}s} \to a(\mathbf{k}, s) \qquad ; \qquad \left(\frac{2\pi\hbar c^2}{\omega_k}\right)^{1/2} \hat{a}_{\mathbf{k}s}^\dagger \to a^\star(\mathbf{k}, s)$$

Hence, recover the classical field for that mode.

Solution to Problem 7.7

(a) The field operator for the vector potential in Eq. (7.45) either creates or destroys a photon, thus changing the photon number by one. Hence the diagonal matrix element of this operator defining the classical field in Eq. (7.47) vanishes if taken between states with a given number of photons.

(b) Define the commutator of the vector potential with the number operator as

$$[\hat{\mathbf{A}}(\mathbf{x}), \, \hat{N}] \equiv i\hat{\mathbf{C}}(\mathbf{x})$$

These are now all hermitian operators, as in Eq. (3.217). It follows as in Eq. (3.220) that there is a generalized uncertainty relation between classical field strength and photon number[6]

$$[\Delta\mathbf{A}(\mathbf{x})]^2 (\Delta N)^2 \geq \tfrac{1}{4}|\langle\hat{\mathbf{C}}(\mathbf{x})\rangle|^2$$

(c) The free field vector potential in the Heisenberg picture in the

---

[5]See the discussion of Bose condensation in chapter 11 of Vol. II.

[6]It is clear from the solution to Prob. 4.10 in [Amore and Walecka (2013)] that Eq. (3.220) as written is in error; it should read $(\Delta A)^2(\Delta B)^2 \geq (1/4)|\langle\hat{C}\rangle|^2$.

Coulomb gauge is given in Prob. (7.8) as

$$\hat{\mathbf{A}}_H(\mathbf{x}, t) \equiv \frac{1}{\sqrt{\Omega}} \sum_{\mathbf{k}} \sum_{s=1}^{2} \left(\frac{2\pi \hbar c^2}{\omega_k} \right)^{1/2} \mathbf{e}_{\mathbf{k}s} \left(\hat{a}_{\mathbf{k}s}\, e^{i\mathbf{k}\cdot\mathbf{x} - i\omega_k t} + \hat{a}^{\dagger}_{\mathbf{k}s}\, e^{-i\mathbf{k}\cdot\mathbf{x} + i\omega_k t} \right)$$

The classical field is then given by Eq. (7.47) as

$$\mathbf{A}(\mathbf{x}, t) = \langle \Psi_H | \hat{\mathbf{A}}_H(\mathbf{x}, t) | \Psi_H \rangle$$

Focus on the vector potential corresponding to a given condensed mode $\{\mathbf{k}, s\}$

$$\hat{\mathbf{A}}_H^C(\mathbf{x}, t) \equiv \frac{1}{\sqrt{\Omega}} \left(\frac{2\pi \hbar c^2}{\omega_k} \right)^{1/2} \mathbf{e}_{\mathbf{k}s} \left(\hat{a}_{\mathbf{k}s}\, e^{i\mathbf{k}\cdot\mathbf{x} - i\omega_k t} + \hat{a}^{\dagger}_{\mathbf{k}s}\, e^{-i\mathbf{k}\cdot\mathbf{x} + i\omega_k t} \right)$$

Consider the matrix element of this expression taken with a state which is a linear superposition of states with different photon number $n_{\mathbf{k}s}$ centered on a value $n_{\mathbf{k}s}^0$ where $n_{\mathbf{k}s}^0 \gg 1$. Now define new operators

$$\hat{\xi}_{\mathbf{k}s} \equiv \frac{\hat{a}_{\mathbf{k}s}}{\sqrt{n_{\mathbf{k}s}^0}} \qquad ; \quad \hat{\xi}^{\dagger}_{\mathbf{k}s} \equiv \frac{\hat{a}^{\dagger}_{\mathbf{k}s}}{\sqrt{n_{\mathbf{k}s}^0}}$$

The commutator of these new operators vanishes for large $n_{\mathbf{k}s}^0$

$$[\hat{\xi}_{\mathbf{k}s}, \hat{\xi}^{\dagger}_{\mathbf{k}s}] = \frac{1}{n_{\mathbf{k}s}^0} \to 0 \qquad ; \quad n_{\mathbf{k}s}^0 \to \infty$$

Since the commutator vanishes, the operators can be treated as c-numbers. We replace[7]

$$\left(\frac{2\pi \hbar c^2 n_{\mathbf{k}s}^0}{\omega_k} \right)^{1/2} \hat{\xi}_{\mathbf{k}s} \to a(\mathbf{k}, s) \qquad ; \quad \left(\frac{2\pi \hbar c^2 n_{\mathbf{k}s}^0}{\omega_k} \right)^{1/2} \hat{\xi}^{\dagger}_{\mathbf{k}s} \to a^{\star}(\mathbf{k}, s)$$

Then

$$\mathbf{A}^C(\mathbf{x}, t) = \frac{1}{\sqrt{\Omega}} \left[a(\mathbf{k}, s) \mathbf{e}_{\mathbf{k},s} e^{i(\mathbf{k}\cdot\mathbf{x} - \omega_k t)} + a(\mathbf{k}, s)^{\star} \mathbf{e}_{\mathbf{k},s} e^{-i(\mathbf{k}\cdot\mathbf{x} - \omega_k t)} \right]$$

This is just the classical field for that mode in Eq. (7.23).

---

[7]It is assumed here that the product $\hbar \omega_k n_{\mathbf{k}s}^0$ remains finite in the classical limit.

Problem 7.8 The Heisenberg picture for the free radiation field is obtained from the Schrödinger picture by

$$|\Psi_H\rangle = e^{i\hat{H}_0 t/\hbar}|\Psi_S(t)\rangle \qquad ; \text{ Heisenberg picture}$$

$$\hat{O}_H(t) = e^{i\hat{H}_0 t/\hbar}\hat{O}_S e^{-i\hat{H}_0 t/\hbar} \qquad \hat{H}_0 \equiv \hat{H}_{\text{rad}}$$

Use Eqs. (7.49) to show the free field operator in the Heisenberg picture is

$$\hat{\mathbf{A}}_H(\mathbf{x}, t) \equiv \frac{1}{\sqrt{\Omega}}\sum_{\mathbf{k}}\sum_{s=1}^{2}\left(\frac{2\pi\hbar c^2}{\omega_k}\right)^{1/2}\mathbf{e}_{\mathbf{k}s}\left(\hat{a}_{\mathbf{k}s}\,e^{i\mathbf{k}\cdot\mathbf{x}-i\omega_k t} + \hat{a}_{\mathbf{k}s}^{\dagger}\,e^{-i\mathbf{k}\cdot\mathbf{x}+i\omega_k t}\right)$$

Compare with the expression in Eq. (7.36).

Solution to Problem 7.8

The vector potential in the Coulomb gauge in the Schrödinger picture is given by Eq. (7.45)

$$\hat{\mathbf{A}}(\mathbf{x}) \equiv \frac{1}{\sqrt{\Omega}}\sum_{\mathbf{k}}\sum_{s=1}^{2}\left(\frac{2\pi\hbar c^2}{\omega_k}\right)^{1/2}\mathbf{e}_{\mathbf{k}s}\left(\hat{a}_{\mathbf{k}s}\,e^{i\mathbf{k}\cdot\mathbf{x}} + \hat{a}_{\mathbf{k}s}^{\dagger}\,e^{-i\mathbf{k}\cdot\mathbf{x}}\right)$$

The commutators required to go to the Heisenberg picture for the free field are given by Eqs. (7.49) as

$$[\hat{H}_{\text{rad}}, \hat{a}_{\mathbf{k}s}] = -\hbar\omega_k\hat{a}_{\mathbf{k}s}$$

$$[\hat{H}_{\text{rad}}, \hat{a}_{\mathbf{k}s}^{\dagger}] = \hbar\omega_k\hat{a}_{\mathbf{k}s}^{\dagger}$$

The calculation for the transformation of the creation and destruction operators is now precisely that of Prob. 3.7, with the result that

$$\hat{\mathbf{A}}_H(\mathbf{x}, t) \equiv \frac{1}{\sqrt{\Omega}}\sum_{\mathbf{k}}\sum_{s=1}^{2}\left(\frac{2\pi\hbar c^2}{\omega_k}\right)^{1/2}\mathbf{e}_{\mathbf{k}s}\left(\hat{a}_{\mathbf{k}s}\,e^{i\mathbf{k}\cdot\mathbf{x}-i\omega_k t} + \hat{a}_{\mathbf{k}s}^{\dagger}\,e^{-i\mathbf{k}\cdot\mathbf{x}+i\omega_k t}\right)$$

The free-field vector potential in the Heisenberg picture now has exactly the same space-time dependence as the classical field in Eq. (7.36).

Problem 7.9 Go to the coordinate representation, and verify the step leading from the first to the second of Eqs. (7.138).

Solution to Problem 7.9

The first of Eqs. (7.138) is

$$\int d^3x \, \langle \phi_f | \hat{\mathbf{S}}(\mathbf{x}) | \phi_i \rangle = \frac{1}{2m_0} \int d^3x \, \langle \phi_f | \hat{\mathbf{p}}_e \delta^{(3)}(\mathbf{x} - \hat{\mathbf{x}}_e) + \delta^{(3)}(\mathbf{x} - \hat{\mathbf{x}}_e) \hat{\mathbf{p}}_e | \phi_i \rangle$$

Make use of a complete set of eigenstates of position satisfying

$$\hat{\mathbf{x}}_e | \boldsymbol{\xi} \rangle = \boldsymbol{\xi} | \boldsymbol{\xi} \rangle$$
$$\int d^3\xi \, | \boldsymbol{\xi} \rangle \langle \boldsymbol{\xi} | = \hat{1}$$

Introduce the completeness relation just after the momentum operator in the first term, and just before the momentum operator in the second term, in the above relation. It then becomes

$$\int d^3x \, \langle \phi_f | \hat{\mathbf{S}}(\mathbf{x}) | \phi_i \rangle = \frac{1}{2m_0} \int d^3x \int d^3\xi$$
$$\times \left[\delta^{(3)}(\mathbf{x} - \boldsymbol{\xi}) \langle \phi_f | \hat{\mathbf{p}}_e | \boldsymbol{\xi} \rangle \langle \boldsymbol{\xi} | \phi_i \rangle + \delta^{(3)}(\mathbf{x} - \boldsymbol{\xi}) \langle \phi_f | \boldsymbol{\xi} \rangle \langle \boldsymbol{\xi} | \hat{\mathbf{p}}_e | \phi_i \rangle \right]$$

Use the delta-functions to do the integral over ξ

$$\int d^3x \, \langle \phi_f | \hat{\mathbf{S}}(\mathbf{x}) | \phi_i \rangle = \frac{1}{2m_0} \int d^3x \, \left[\langle \phi_f | \hat{\mathbf{p}}_e | \mathbf{x} \rangle \langle \mathbf{x} | \phi_i \rangle + \langle \phi_f | \mathbf{x} \rangle \langle \mathbf{x} | \hat{\mathbf{p}}_e | \phi_i \rangle \right]$$

Now use the completeness of the eigenstates of position once again to obtain

$$\int d^3x \, \langle \phi_f | \hat{\mathbf{S}}(\mathbf{x}) | \phi_i \rangle = \frac{1}{2m_0} \left[\langle \phi_f | \hat{\mathbf{p}}_e | \phi_i \rangle + \langle \phi_f | \hat{\mathbf{p}}_e | \phi_i \rangle \right]$$
$$= \frac{1}{m_0} \langle \phi_f | \hat{\mathbf{p}}_e | \phi_i \rangle$$

This is the second of Eqs. (7.138).

Problem 7.10 (a) Show that the $1s \to 2p$ transition in H-like atoms makes up a fraction of $2^{13}/3^9 = 0.416$ of the dipole sum rule;

(b) Take out a mean excitation energy $\bar{\varepsilon}$, and use closure to re-write the dipole sum rule as

$$\sum_n (\varepsilon_n - \varepsilon_i) |\langle \phi_n | \mathbf{x} | \phi_i \rangle|^2 = \bar{\varepsilon} \, \langle \phi_i | r^2 | \phi_i \rangle = \frac{3\hbar^2}{2m_0}$$

Evaluate this expression for H-like atoms, and show that $\bar{\varepsilon} = |\varepsilon_{1s}|$, which is the magnitude of the binding energy of the ground state.

Solution to Problem 7.10

(a) The dipole sum rule is given in Eq. (7.186)

$$\sum_n (\varepsilon_n - \varepsilon_i) \left| \langle \phi_n | \mathbf{x} | \phi_i \rangle \right|^2 = \frac{3\hbar^2}{2m_0}$$

The contribution to the sum rule from the $1s \rightarrow 2p$ transition in H-like atoms is

$$D_{2p\leftarrow 1s} = \sum_{m_f} (\varepsilon_{2p} - \varepsilon_{1s}) \left| \langle \phi_{2p\, m_f} | \mathbf{x} | \phi_{1s\, 0} \rangle \right|^2$$

It follows from Eqs. (7.144) and (7.152) that

$$D_{2p\leftarrow 1s} = \frac{4\pi}{3} \hbar \omega_k \left| \int_0^\infty r^2 dr\, R_{21}(r) r R_{10}(r) \right|^2 \sum_{m_f} \sum_m$$

$$\times \left| \int d\Omega_x\, Y_{1,m_f}^\star(\Omega_x) Y_{1,m}(\Omega_x) Y_{00}(\Omega_x) \right|^2$$

This is evaluated as follows:

- The excitation energy is given in Eq. (7.169)

$$\hbar \omega_k = \varepsilon_{2p} - \varepsilon_{1s} = \frac{3}{8}(Z\alpha)^2 m_0 c^2$$

- The required radial integral is given in Eq. (7.166)

$$\int_0^\infty r^2 dr\, R_{2p}(r) r R_{1s}(r) = \frac{1}{\sqrt{6}} \frac{2^8}{3^4} \frac{a_0}{Z}$$

- The angular contribution follows from Eq. (7.167)

$$\frac{4\pi}{3} \sum_{m_f} \sum_m \left| \int d\Omega_x\, Y_{1,m_f}^\star(\Omega_x) Y_{1,m}(\Omega_x) Y_{0,0}(\Omega_x) \right|^2 = 1$$

A combination of these results gives

$$D_{2p\leftarrow 1s} = \left[\frac{3}{8}(Z\alpha)^2 m_0 c^2 \right] \left[\frac{1}{\sqrt{6}} \frac{2^8}{3^4} \frac{a_0}{Z} \right]^2$$

With $a_0 = \hbar/\alpha m_0 c$, this is

$$D_{2p\leftarrow 1s} = \frac{2^{13}}{3^9} \frac{3\hbar^2}{2m_0}$$

which is the stated answer.

(b) If we replace the excitation energy in the dipole sum rule by a mean value, completeness can be invoked to write

$$\frac{3\hbar^2}{2m_0} = \sum_n (\varepsilon_n - \varepsilon_i) |\langle \phi_n | \mathbf{x} | \phi_i \rangle|^2$$

$$= \bar{\varepsilon} \sum_j \sum_n \langle \phi_i | \mathbf{x}_j | \phi_n \rangle \langle \phi_n | \mathbf{x}_j | \phi_i \rangle = \bar{\varepsilon} \sum_j \langle \phi_i | \mathbf{x}_j \mathbf{x}_j | \phi_i \rangle$$

$$= \bar{\varepsilon} \langle \phi_i | \mathbf{x}^2 | \phi_i \rangle = \bar{\varepsilon} \langle \phi_i | r^2 | \phi_i \rangle$$

To proceed, we need the ground-state expectation value of r^2 for H-like atoms. This is computed with the wave function in Prob. 2.22 as

$$\langle \phi_{1s} | r^2 | \phi_{1s} \rangle = 4 \left(\frac{Z}{a_0} \right)^3 \int_0^\infty r^4 dr\, e^{-2Zr/a_0}$$

$$= \frac{1}{8} \left(\frac{a_0}{Z} \right)^2 \int_0^\infty t^4 e^{-t}\, dt = 3 \left(\frac{a_0}{Z} \right)^2$$

The mean excitation energy is then

$$\bar{\varepsilon} = \frac{1}{\langle \phi_{1s} | r^2 | \phi_{1s} \rangle} \frac{3\hbar^2}{2m_0} = \left(\frac{Z}{a_0} \right)^2 \frac{\hbar^2}{2m_0}$$

With $a_0 = \hbar/\alpha m_0 c$, this reduces to

$$\bar{\varepsilon} = \frac{1}{2} \left(Z^2 \alpha^2 \right) m_0 c^2 = |\varepsilon_{1s}|$$

where $|\varepsilon_{1s}|$ is the magnitude of the ground-state binding energy.

Problem 7.11 A particle moves in an isotropic three-dimensional harmonic oscillator. Show the dipole sum rule is saturated by transitions which go up one oscillator spacing.[8]

Solution to Problem 7.11

As in Prob. 2.27, we treat the three-dimensional oscillator in factored form, where the eigenstates are the product of three one-dimensional oscillators. The l.h.s. of the dipole sum rule then takes the form

$$\text{l.h.s.} = \sum_{n_x} \sum_{n_y} \sum_{n_z} \hbar\omega_0 (n_x + n_y + n_z) |\langle n_x n_y n_z | \mathbf{x} | 000 \rangle|^2$$

---

[8]Recall Prob. 2.27.

The matrix element is calculated exactly as in Eq. (6.33)

$$\langle n_x n_y n_z | \mathbf{x} | 000 \rangle = \langle n_x n_y n_z | x\mathbf{e}_x + y\mathbf{e}_y + z\mathbf{e}_z | 000 \rangle$$

$$= \left(\frac{\hbar}{2m_0\omega_0} \right)^{1/2} \left(\mathbf{e}_x\, \delta_{n_x,1}\, \delta_{n_y,0}\, \delta_{n_z,0} + \mathbf{e}_y\, \delta_{n_x,0}\, \delta_{n_y,1}\, \delta_{n_z,0} + \mathbf{e}_z\, \delta_{n_x,0}\, \delta_{n_y,0}\, \delta_{n_z,1} \right)$$

Note that each term only goes up one oscillator spacing. The l.h.s. of the dipole sum rule then becomes

$$\text{l.h.s.} = \hbar\omega_0 \left(\frac{\hbar}{2m_0\omega_0} \right) \left(\mathbf{e}_x^2 + \mathbf{e}_y^2 + \mathbf{e}_z^2 \right)$$

$$= \frac{3\hbar^2}{2m_0}$$

which saturates the sum rule.

Problem 7.12 Given the mean life τ as a time interval Δt, and using the half-width γ from Fig. 7.14 in the text as a measure of the spread in the photon energy ΔE, deduce an uncertainty relation for photoemission[9]

$$\Delta E \Delta t = \hbar \qquad ; \text{ photoemission}$$

Solution to Problem 7.12

The mean life is identified through Eq. (7.222)

$$|c_1(t)|^2 = e^{-\gamma t} \equiv e^{-t/\tau}$$

Therefore

$$\tau = \frac{1}{\gamma}$$

From Fig. 7.14 in the text, the half-width Γ of the line, in the energy difference $\hbar\nu = \hbar(\omega - \omega_{12})$, is

$$\Gamma = \hbar\gamma$$

Hence if we define an energy uncertainty of $\Delta E \equiv \Gamma$, and take a time interval of $\Delta t \equiv \tau$, there is an "uncertainty relation"

$$\Delta E \Delta t = \frac{\hbar}{\tau} \tau = \hbar$$

---

[9] *Hint*: Recall Eq. (7.222).

Problem 7.13 Suppose one truncates the integral over the photon angular frequencies to an *asymmetric* region of integration in Eq. (7.217).

(a) Show the term left over from the first integral in Eq. (7.219) would add a small imaginary part to γ, so that $\gamma \to \gamma_R + i\gamma_I$;

(b) Show that the imaginary part $i\gamma_I$ would give rise to a small real energy shift of the levels. (But so do the electromagnetic self-energies of the levels, arising from emission and absorption of photons, which have been neglected here.)

Solution to Problem 7.13

(b) Let us do part (b) first. Suppose γ has a small imaginary part so that

$$\gamma \to \gamma_R + i\gamma_I$$

Then in Eq. (7.211), one has

$$\frac{\gamma}{2} + i(\omega_{12} - w) \to \frac{\gamma_R}{2} + i(\tilde{\omega}_{12} - w) \qquad ; \ \tilde{\omega}_{12} \equiv \omega_{12} + \frac{\gamma_I}{2}$$

This is equivalent to a small real energy shift of the levels.

(a) Call γ_R the result in Eq. (7.221). Then if one truncates the integral over the photon angular frequencies to an asymmetric region of integration in Eq. (7.217), one has with the aid of Eq. (7.219)

$$\gamma \to \gamma_R + i\gamma_I$$

$$\gamma_I \equiv \frac{\gamma_R}{\pi} \int_{\nu_0}^{\nu_0 + \delta\nu_0} d\nu \left(\frac{e^{-i\nu t} - 1}{\nu} \right)$$

Assume that $|\delta\nu_0|/\nu_0 \ll 1$, $|\gamma|/\nu_0 \ll 1$, and as in Eq. (7.220), that $(\delta\nu_0)t \gg 1$. Then

$$\gamma_I \approx \frac{\gamma_R}{\pi} \frac{1}{\nu_0} \int_{\nu_0}^{\nu_0 + \delta\nu_0} d\nu \left(e^{-i\nu t} - 1 \right)$$

$$\approx -\frac{\gamma_R}{\pi} \frac{\delta\nu_0}{\nu_0}$$

Hence γ then acquires a small imaginary part.

Problem 7.14 (a) Establish the representation of the Dirac delta-function in Eq. (7.228);

(b) Use the result in (a) to justify Eqs. (7.201) and (6.176) for small, but finite, γ.

Solution to Problem 7.14

The distribution in photon energies $\hbar\omega$ in the Wigner-Weisskopf theory of the line shape for the transition $1 \to 2$, where 2 is the ground state, is given by Eq. (7.226)

$$N(\nu) = \frac{\gamma}{2\pi} \frac{1}{\nu^2 + \gamma^2/4} \qquad ; \nu = \omega - \omega_{12}$$

Here γ is the transition rate, or inverse lifetime, given in Eq. (7.221)

$$\gamma = \frac{2\pi}{\hbar^2} \left[\int \sum_s |\langle \phi_1 | \hat{H}_1 | \phi_2; \mathbf{k}s \rangle|^2 \left(\frac{\partial n_f}{\partial \nu} \right) \right]_{\varepsilon_1 = \varepsilon_2 + \hbar\omega}$$

(a) As $\gamma \to 0$, the function $N(\nu)$ has the two key features of a Dirac delta-function

(1) As $\gamma \to 0$, $N(\nu)$ vanishes unless $\nu = 0$, in which case it is infinite

$$N(\nu) \to 0 \qquad\qquad ; \gamma \to 0 \qquad ; \nu \neq 0$$
$$N(\nu) = \frac{2}{\pi\gamma} \to \infty \qquad ; \gamma \to 0 \qquad ; \nu = 0$$

(2) The integral of $N(\nu)$ over ν is finite for all γ. With $2\nu/\gamma \equiv x$, one has

$$\int_{-\infty}^{\infty} N(\nu)\, d\nu = \frac{\gamma}{2\pi} \int_{-\infty}^{\infty} d\nu \frac{1}{\nu^2 + \gamma^2/4} = \frac{1}{\pi} \int_{-\infty}^{\infty} \frac{dx}{1 + x^2}$$
$$= 1$$

We conclude that

$$\mathrm{Lim}_{\gamma \to 0}\, N(\nu) = \delta(\nu)$$

(b) In computing the photoabsorption cross section, we had the sharply-peaked factor $\delta(\hbar\omega - \varepsilon_1 + \varepsilon_2)$. Wigner-Weisskopf theory suggests this should be the actual line shape $N(\nu)/\hbar$. Then as the half-width $\gamma \to 0$,

$$\frac{1}{\hbar} \mathrm{Lim}_{\gamma \to 0}\, N(\nu) = \frac{1}{\hbar} \delta(\omega - \omega_{12})$$
$$= \delta(\hbar\omega - \varepsilon_1 + \varepsilon_2)$$

PART 3
Relativistic Quantum Field Theory

Chapter 8

Discrete Symmetries

Problem 8.1 Give an argument that Eq. (8.20) represents the most general state vector with angular momentum L in the C-M system formed from a (π^+, π^-) pair.

Solution to Problem 8.1

Consider the following state

$$|ELM\rangle = \int d^3k \, a_E(k) Y_{LM}(\Omega_k) \, a_{\mathbf{k}}^\dagger b_{-\mathbf{k}}^\dagger |0\rangle$$

This state contains a π^+ with momentum $\hbar \mathbf{k}$ created by $a_{\mathbf{k}}^\dagger$ and a π^- with momentum $-\hbar \mathbf{k}$ created by $b_{\mathbf{k}}^\dagger$. The total momentum of the state is $\mathbf{P} = 0$. We then sum over all values of $\mathbf{k}$ with an arbitrary weighting of $a_E(k)$, depending on the energy eigenvalue. Although one could add additional pairs to the state, created, for example, through vacuum polarization, this is the most general state with a single (π^+, π^-) pair and zero total momentum.

As to the angular momentum properties, consider the behavior of this state under rotation. Let the rotation operator $\hat{R}_\omega$ act on the state

$$\hat{R}_\omega |ELM\rangle = \int d^3k \, a_E(k) Y_{LM}(\Omega_k) \, \hat{R}_\omega \, a_{\mathbf{k}}^\dagger b_{-\mathbf{k}}^\dagger |0\rangle$$

$$= \int d^3k \, a_E(k) Y_{LM}(\Omega_k) \, a_{\mathbf{k}'}^\dagger b_{-\mathbf{k}'}^\dagger |0\rangle$$

Here $\mathbf{k}'$ points in the rotated direction. Now change dummy integration variables from $\mathbf{k}$ to the rotated values $\mathbf{k}'$, with $k = k'$ and $d^3k = d^3k'$, and use the behavior of the spherical harmonics under rotation given in

199

Eq. (4.1.4) in [Edmonds (1974)][1]

$$Y_{LM}(\Omega_k) = \sum_{M'} \mathcal{D}^L_{M'M}(\omega)\, Y_{LM'}(\Omega_{k'})$$

Hence

$$\hat{R}_\omega |ELM\rangle = \sum_{M'} \mathcal{D}^L_{M'M}(\omega) \int d^3k'\, a_E(k') Y_{LM'}(\Omega_{k'})\, a^\dagger_{\mathbf{k}'} b^\dagger_{-\mathbf{k}'} |0\rangle$$

$$= \sum_{M'} \mathcal{D}^L_{M'M}(\omega) |ELM'\rangle$$

This demonstrates that the constructed state vector is an appropriate eigenstate of angular momentum.

Problem 8.2 (a) Review Probs. 9.6 and 9.7 in Vol. I. Show that the following η_λ are two-component eigenstates of the helicity operator $\boldsymbol{\sigma}\cdot(\mathbf{k}/k)$ with eigenvalues $\lambda = \pm 1$

$$\eta_{+1} = \begin{pmatrix} e^{-i\phi_k/2}\cos\theta_k/2 \\ e^{i\phi_k/2}\sin\theta_k/2 \end{pmatrix} \qquad ; \; \eta_{-1} = \begin{pmatrix} -e^{-i\phi_k/2}\sin\theta_k/2 \\ e^{i\phi_k/2}\cos\theta_k/2 \end{pmatrix}$$

(b) Now derive the results in Table 8.1 in the text.[2]

Solution to Problem 8.2

(a) Write the unit vector $\mathbf{k}/k$ in spherical coordinates as

$$\mathbf{k}/k = (\sin\theta\cos\phi,\ \sin\theta\sin\phi,\ \cos\theta)$$

---

[1]Rather than re-deriving this result, let us simply check a couple of cases to see that we have the correct interpretation and signs. Let $L = 1$. Suppose the state vector is rotated by an angle γ about the z-axis. The operator that does this for us is $\exp(-i\gamma\hat{J}_z)$. Then

$$\sum_{M'} \mathcal{D}^1_{M'M}(0,0,-\gamma) Y_{1M'}(\theta_k,\phi_k+\gamma) = Y_{1M}(\theta_k,\phi_k)$$

Suppose $\phi_k = 0$ and the state vector is rotated by an angle β about the y-axis. The operator that produces this is $\exp(-i\beta\hat{J}_y)$. Then

$$\sum_{M'} \mathcal{D}^1_{M'M}(0,-\beta,0) Y_{1M'}(\theta_k+\beta,0) = Y_{1M}(\theta_k,0)$$

where use has been made of $d^1_{M'M}(-\beta)$ from [Edmonds (1974)].

[2]Note that the overall phase of the last eigenfunction is a matter of convention.

where $(\theta, \phi) = (\theta_k, \phi_k)$. The 2×2 matrix $\boldsymbol{\sigma} \cdot (\mathbf{k}/k)$ is then given by

$$\boldsymbol{\sigma} \cdot \frac{\mathbf{k}}{k} = \begin{pmatrix} \cos\theta & e^{-i\phi}\sin\theta \\ e^{i\phi}\sin\theta & -\cos\theta \end{pmatrix}$$

Now apply this to the two-component spinor η_{+1}

$$\boldsymbol{\sigma} \cdot \frac{\mathbf{k}}{k}\eta_{+1} = \begin{pmatrix} e^{-i\phi/2}(\cos\theta\cos\theta/2 + \sin\theta\sin\theta/2) \\ e^{+i\phi/2}(\sin\theta\cos\theta/2 - \cos\theta\sin\theta/2) \end{pmatrix}$$

Use the half-angle formulae to re-write this as

$$\boldsymbol{\sigma} \cdot \frac{\mathbf{k}}{k}\eta_{+1} = \begin{pmatrix} e^{-i\phi/2}(\cos^3\theta/2 - \sin^2\theta/2\cos\theta/2 + 2\sin^2\theta/2\cos\theta/2) \\ e^{+i\phi/2}(2\sin\theta/2\cos^2\theta/2 - \cos^2\theta/2\sin\theta/2 + \sin^3\theta/2) \end{pmatrix}$$
$$= \begin{pmatrix} e^{-i\phi/2}\cos\theta/2 \\ e^{i\phi/2}\sin\theta/2 \end{pmatrix} = \eta_{+1}$$

In a similar fashion, one has

$$\boldsymbol{\sigma} \cdot \frac{\mathbf{k}}{k}\eta_{-1} = \begin{pmatrix} e^{-i\phi/2}(-\cos\theta\sin\theta/2 + \sin\theta\cos\theta/2) \\ e^{+i\phi/2}(-\sin\theta\sin\theta/2 - \cos\theta\cos\theta/2) \end{pmatrix}$$

Re-written in terms of half-angle formulae, this gives

$$\boldsymbol{\sigma} \cdot \frac{\mathbf{k}}{k}\eta_{-1} = \begin{pmatrix} e^{-i\phi/2}(-\cos^2\theta/2\sin\theta/2 + \sin^3\theta/2 + 2\sin\theta/2\cos^2\theta/2) \\ e^{+i\phi/2}(-2\sin^2\theta/2\cos\theta/2 - \cos^3\theta/2 + \sin^2\theta/2\cos\theta/2) \end{pmatrix}$$
$$= \begin{pmatrix} e^{-i\phi/2}\sin\theta/2 \\ -e^{i\phi/2}\cos\theta/2 \end{pmatrix} = -\eta_{-1}$$

This establishes the results in part (a).[3]

(b) Here, we first need some kinematics. From Eqs. (8.54), we have

$$(r+s)^2 = (\cos\chi/2 + \sin\chi/2)^2 = 1 + \sin\chi = \frac{E_k + m_0c^2}{E_k}$$
$$(r-s)^2 = (\cos\chi/2 - \sin\chi/2)^2 = 1 - \sin\chi = \frac{E_k - m_0c^2}{E_k}$$

---

[3]Note the spinors $\eta_{\pm 1}$ are just the appropriate *rotations* of the elementary two-component spinors with spin-up and spin-down along the z-axis; this sets their overall phase.

Hence

$$\frac{1}{\sqrt{2}}(r+s) = \left[\frac{E_k + m_0 c^2}{2E_k}\right]^{1/2}$$

$$\frac{1}{\sqrt{2}}(r-s) = \left[\frac{E_k - m_0 c^2}{2E_k}\right]^{1/2} = \frac{\hbar k c}{[2E_k(E_k + m_0 c^2)]^{1/2}}$$

Now from Prob. 9.6 in Vol. I, the positive-energy Dirac spinors are

$$u_\lambda(\mathbf{k}) = \left[\frac{E_k + m_0 c^2}{2E_k}\right]^{1/2} \left(\begin{matrix} \eta_\lambda \\ [\hbar c \boldsymbol{\sigma} \cdot \mathbf{k}/(E_k + m_0 c^2)] \, \eta_\lambda \end{matrix}\right)$$

where η_λ are two-component eigenstates of helicity. With the use of the results in part (a), we then reproduce the first two columns of Table 8.1 in the text.

From Prob. 9.6 in Vol. I, the negative-energy Dirac spinors are

$$v_\lambda(-\mathbf{k}) = \left[\frac{E_k + m_0 c^2}{2E_k}\right]^{1/2} \left(\begin{matrix} [\hbar c \boldsymbol{\sigma} \cdot \mathbf{k}/(E_k + m_0 c^2)] \, \eta_\lambda \\ \eta_\lambda \end{matrix}\right)$$

where η_λ are two-component eigenstates of helicity with respect to $-\mathbf{k}$. With the use of the results from part (a), and

$$\boldsymbol{\sigma} \cdot (-\mathbf{k}) \, \eta_\lambda = -\boldsymbol{\sigma} \cdot \mathbf{k} \, \eta_\lambda$$

which simply interchanges the role of the helicity eigenstates, we then reproduce the last two columns in Table 8.1 in the text.[4]

Problem 8.3 (a) Verify Eqs. (8.65);
(b) Verify Eq. (8.156);
(c) Verify the statement made after Eq. (8.141).

---

[4]The overall phase of the last eigenfunction is a matter of convention, chosen so the wave functions in Table 8.1 are related by the simple substitution $s \rightleftharpoons -s$. (Note the spinors are double-valued, and an additional rotation $\phi_k \rightarrow \phi_k + 2\pi$ produces this minus sign.)

Solution to Problem 8.3

(a) Let us calculate $\sqrt{2}\,v(-\mathbf{k}\uparrow)^\dagger\gamma_2$ from Eqs. (8.53) and Table 8.1

$$\sqrt{2}\,v(-\mathbf{k}\uparrow)^\dagger\gamma_2 = [-b(r-s),\,a(r-s),\,b(r+s),\,-a(r+s)]\begin{bmatrix} & & & -1 \\ & & 1 & \\ & 1 & & \\ -1 & & & \end{bmatrix}$$

$$= [a(r+s),\,b(r+s),\,a(r-s),\,b(r-s)]$$
$$= \sqrt{2}\,u(\mathbf{k}\uparrow)^T$$

In a similar fashion

$$\sqrt{2}\,v(-\mathbf{k}\downarrow)^\dagger\gamma_2 = [a^\star(r-s),\,b^\star(r-s),\,a^\star(r+s),\,b^\star(r+s)]\begin{bmatrix} & & & -1 \\ & & 1 & \\ & 1 & & \\ -1 & & & \end{bmatrix}$$

$$= [-b^\star(r+s),\,a^\star(r+s),\,b^\star(r-s),\,-a^\star(r-s)]$$
$$= \sqrt{2}\,u(\mathbf{k}\downarrow)^T$$

The first set of Eqs. (8.65) constitutes the component form of these relations.

The second set of Eqs. (8.65) can be obtained from the above through the substitutions $s \to -s$, $v(-\mathbf{k}\downarrow) \rightleftharpoons u(\mathbf{k}\uparrow)$, and $v(-\mathbf{k}\uparrow) \rightleftharpoons u(\mathbf{k}\downarrow)$.

(b) With S as defined in Eqs. (8.153), the Dirac field transforms according to Eq. (8.155)

$$\hat{U}(L)\hat{\psi}(x)\hat{U}(L)^{-1} = S\hat{\psi}(x') \qquad ; \; x'_\mu = a_{\mu\nu}(-v)x_\nu$$

Make use of the properties of S in Eqs. (8.154)

$$\gamma_4 S^\dagger \gamma_4 = S^{-1}$$
$$S\gamma_\mu S^{-1} = a_{\mu\nu}(-v)\gamma_\nu$$

The orthogonality relation $a_{\mu\nu}a_{\mu\lambda} = \delta_{\nu\lambda}$ allows the second relation to be inverted into

$$a_{\mu\lambda}(-v)\gamma_\mu = S^{-1}\gamma_\lambda S$$

The Lorentz transformation property of the Dirac current in Eq. (8.80) then follows according to

$$\hat{U}(L)\hat{\bar{\psi}}(x)\gamma_\mu\hat{\psi}(x)\hat{U}(L)^{-1} = \hat{U}(L)\hat{\psi}^\dagger(x)\hat{U}(L)^{-1}\gamma_4\gamma_\mu\hat{U}(L)\hat{\psi}(x)\hat{U}(L)^{-1}$$
$$= \hat{\psi}^\dagger(x')S^\dagger\gamma_4\gamma_\mu S\hat{\psi}(x')$$
$$= \hat{\bar{\psi}}(x')S^{-1}\gamma_\mu S\hat{\psi}(x')$$
$$= a_{\nu\mu}(-v)\hat{\bar{\psi}}(x')\gamma_\nu\hat{\psi}(x')$$

Thus

$$\hat{U}(L)\hat{j}_\mu(x)\hat{U}(L)^{-1} = a_{\nu\mu}(-v)\hat{j}_\nu(x')$$

which is just Eq. (8.156).

(c) In the Coulomb gauge, The electric field is related to the vector potential by Eqs. (7.18)

$$\mathbf{E} = -\frac{1}{c}\frac{\partial\mathbf{A}}{\partial t}$$

Hence for the quantized radiation field, from Eq. (8.101),

$$\mathbf{E} = i\sum_{k}\sum_{s=1}^{2}(2\pi\hbar\omega_k)^{1/2}\left(a_{\mathbf{k}s}\mathbf{e}_{\mathbf{k}s}e^{ik\cdot x} - a_{\mathbf{k}s}^\dagger\mathbf{e}_{\mathbf{k}s}e^{-ik\cdot x}\right)$$

Here $a_{\mathbf{k}s}^\dagger$ creates a photon with polarization vector $\mathbf{e}_{\mathbf{k}s}$.

The polarization vectors for $-\mathbf{k}$ and $\mathbf{k}$ are related in Fig. 8.8 in the text

$$\mathbf{e}_{-\mathbf{k},1} = -\mathbf{e}_{\mathbf{k},1} \qquad ; \mathbf{e}_{-\mathbf{k},2} = \mathbf{e}_{\mathbf{k},2}$$

From Eq. (8.141), the two-photon state with $J^\pi = 0^-$ in Table 8.5 in the text can be written

$$|\mathbf{k}, -\mathbf{k}; 0^-\rangle = \frac{i}{\sqrt{2}}\left(a_{\mathbf{k}1}^\dagger a_{-\mathbf{k}2}^\dagger + a_{\mathbf{k}2}^\dagger a_{-\mathbf{k}1}^\dagger\right)|0\rangle$$

Since the polarization vectors $\mathbf{e}_{\mathbf{k}1}$ and $\mathbf{e}_{-\mathbf{k}2}$ are orthogonal to each other, as are $\mathbf{e}_{\mathbf{k}2}$ and $\mathbf{e}_{-\mathbf{k}1}$, this represents a state where the planes of polarization of the $\mathbf{E}$-vectors of the two photons are perpendicular.

Problem 8.4 (a) Use the orthonormality of the spinors $\bar{\mathcal{U}}(\mathbf{k}\lambda)\mathcal{U}(\mathbf{k}\lambda') = \delta_{\lambda\lambda'}$ to prove from Eq. (8.152) that the 2×2 matrix $\alpha_{\lambda\lambda'}$ is unitary

$$\sum_{\lambda''}\alpha_{\lambda\lambda''}\alpha_{\lambda'\lambda''}^\star = \delta_{\lambda\lambda'}$$

(b) Then show Eq. (8.155) follows from Eqs. (8.150)–(8.154).

Solution to Problem 8.4

(a) Equation (8.152) states that

$$S^{-1}\mathcal{U}(\mathbf{k}\lambda) = \sum_{\lambda'} \alpha_{\lambda\lambda'} \mathcal{U}(\mathbf{k}'\lambda')$$

where S is the 4×4 matrix in Eq. (8.153) corresponding to the Lorentz transformation $k'_\mu = a_{\mu\nu}(-v)k_\nu$, and $\mathcal{U}(\mathbf{k}\lambda)$ is the Dirac spinor with invariant norm satisfying

$$\bar{\mathcal{U}}(\mathbf{k}\lambda)\mathcal{U}(\mathbf{k}\lambda') = \delta_{\lambda\lambda'}$$

From Eqs. (8.154), the matrix S satisfies

$$\gamma_4 S^\dagger \gamma_4 = S^{-1}$$
$$(S^{-1})^\dagger = \gamma_4 S \gamma_4$$

Now multiply Eq. (8.152) by γ_4, and then by the adjoint of Eq. (8.152)

$$\sum_{\lambda'} \alpha_{\lambda\lambda'} \sum_{\lambda''} \alpha^\star_{\underline{\lambda}\lambda''} \mathcal{U}^\dagger(\mathbf{k}'\lambda'')\gamma_4 \mathcal{U}(\mathbf{k}'\lambda') = \mathcal{U}^\dagger(\mathbf{k}\underline{\lambda})(S^{-1})^\dagger \gamma_4 S^{-1}\mathcal{U}(\mathbf{k}\,\lambda)$$

Use the orthonormality of the Dirac spinors on the l.h.s.

$$\bar{\mathcal{U}}(\mathbf{k}'\lambda'')\mathcal{U}(\mathbf{k}'\lambda') = \delta_{\lambda'\lambda''}$$

and use the above properties of S on the r.h.s.

$$\mathcal{U}^\dagger(\mathbf{k}\underline{\lambda})(S^{-1})^\dagger \gamma_4 S^{-1}\mathcal{U}(\mathbf{k}\,\lambda) = \mathcal{U}^\dagger(\mathbf{k}\underline{\lambda})\gamma_4 S S^{-1}\mathcal{U}(\mathbf{k}\,\lambda) = \bar{\mathcal{U}}(\mathbf{k}\underline{\lambda})\mathcal{U}(\mathbf{k}\,\lambda)$$
$$= \delta_{\lambda\underline{\lambda}}$$

A combination of these results gives[5]

$$\sum_{\lambda''} \alpha_{\lambda\lambda''}\alpha^\star_{\lambda'\lambda''} = \delta_{\lambda\lambda'}$$

which is the desired unitarity relation satisfied by the coefficients $\alpha_{\lambda\lambda'}$.

(b) The Dirac field is given in Eq. (8.150)

$$\hat{\psi}(x) = \frac{\sqrt{2M}}{(2\pi)^3} \int d^4k\, \delta(k^2 + M^2)\theta(k_0)$$
$$\times \sum_\lambda \left[\tilde{a}_{\mathbf{k}\lambda}\mathcal{U}(\mathbf{k}\lambda)e^{ik\cdot x} + \tilde{b}^\dagger_{\mathbf{k}\lambda}\mathcal{V}(-\mathbf{k}\lambda)e^{-ik\cdot x} \right]$$

---

[5] We re-label $\underline{\lambda} \to \lambda'$.

Under the unitary Lorentz transformation $\hat{U}(L)$, the creation and destruction operators transform according to Eqs. (8.151)

$$\hat{U}(L)\tilde{a}_{\mathbf{k}\lambda}^{\dagger}\hat{U}(L)^{-1} = \sum_{\lambda'} \alpha_{\lambda\lambda'}\, \tilde{a}_{\mathbf{k}'\lambda'}^{\dagger}\,, \qquad ; \; \hat{U}(L)\tilde{b}_{\mathbf{k}\lambda}^{\dagger}\hat{U}(L)^{-1} = \sum_{\lambda'} \alpha_{\lambda\lambda'}^{\star}\, \tilde{b}_{\mathbf{k}'\lambda'}^{\dagger}$$

It follows that

$$\hat{U}(L)\hat{\psi}(x)\hat{U}(L)^{-1} = \frac{\sqrt{2M}}{(2\pi)^3} \int d^4k\, \delta(k^2 + M^2)\theta(k_0)$$
$$\times \sum_{\lambda}\sum_{\lambda'}\left[\alpha_{\lambda\lambda'}^{\star}\mathcal{U}(\mathbf{k}\lambda)\tilde{a}_{\mathbf{k}'\lambda'}e^{ik\cdot x} + \alpha_{\lambda\lambda'}^{\star}\mathcal{V}(-\mathbf{k}\lambda)\tilde{b}_{\mathbf{k}'\lambda'}^{\dagger}e^{-ik\cdot x}\right]$$

Now use the results in part (a) to express the Dirac spinors in terms of the Lorentz-transformed quantities

$$\mathcal{U}(\mathbf{k}\lambda) = S\sum_{\lambda''}\alpha_{\lambda\lambda''}\mathcal{U}(\mathbf{k}'\lambda'') \qquad ; \; \mathcal{V}(-\mathbf{k}\lambda) = S\sum_{\lambda''}\alpha_{\lambda\lambda''}\mathcal{V}(-\mathbf{k}'\lambda'')$$

Then use the unitarity of the transformation coefficients from part (a)

$$\sum_{\lambda}\alpha_{\lambda\lambda''}\alpha_{\lambda\lambda'}^{\star} = \delta_{\lambda'\lambda''}$$

This gives Eqs. (8.155)[6]

$$\hat{U}(L)\hat{\psi}(x)\hat{U}(L)^{-1} = S\hat{\psi}(x') \qquad ; \; x_{\mu}' = a_{\mu\nu}(-v)x_{\nu}$$

Problem 8.5 The following interaction lagrangian density

$$\hat{\mathcal{L}}_I(x) = ig\hat{\bar{\psi}}\gamma_5\boldsymbol{\tau}\hat{\psi}\cdot\hat{\boldsymbol{\phi}} + if\hat{\bar{\psi}}\gamma_5\gamma_\mu\boldsymbol{\tau}\hat{\psi}\cdot\frac{\partial\hat{\boldsymbol{\phi}}}{\partial x_\mu}$$

represents the most general Yukawa coupling between a nucleon and isovector pseudoscalar pion which is

- Lorentz invariant
- hermitian
- invariant under $\hat{P}$
- invariant under isospin rotations

Show that this $\hat{\mathcal{L}}_I(x)$ is *separately* invariant under $\hat{C}$ and $\hat{T}$.[7]

---

[6]Recall $k \cdot x = k' \cdot x'$.

[7]The charged field is written in terms of the hermitian fields as $\sqrt{2}\,\hat{\phi}^{\star} = \hat{\phi}_1 + i\hat{\phi}_2$. [Note that (f, g) are real.]

Solution to Problem 8.5

This Yukawa-interaction lagrangian density is Lorentz invariant, hermitian, and, since $\bar{\psi}\boldsymbol{\tau}\psi$ transforms as an isovector, invariant under isospin rotations. To investigate the discrete symmetries in the interaction picture, we need the transformation properties of the pion fields in Eqs. (8.1). The charged fields are related to the hermitian components by[8]

$$\hat{\phi}^{\star} = \frac{1}{\sqrt{2}}(\hat{\phi}_1 + i\hat{\phi}_2) \qquad ; \ \hat{\phi} = \frac{1}{\sqrt{2}}(\hat{\phi}_1 - i\hat{\phi}_2) \qquad ; \ \hat{\phi}_3 = \hat{\phi}_3$$

It follows from Eqs. (8.15), (8.26), and (8.44) that the behavior of the hermitian components under the discrete symmetries is

$$\hat{P}\hat{\phi}_i(\mathbf{x},t)\hat{P}^{-1} = \eta_{p_\pi}\hat{\phi}_i(-\mathbf{x},t)$$
$$\hat{C}\hat{\phi}_{1,3}(\mathbf{x},t)\hat{C}^{-1} = \eta_{c_\pi}\hat{\phi}_{1,3}(\mathbf{x},t) \qquad ; \ \hat{C}\hat{\phi}_2(\mathbf{x},t)\hat{C}^{-1} = -\eta_{c_\pi}\hat{\phi}_2(\mathbf{x},t)$$
$$\hat{T}\hat{\phi}_{1,3}(\mathbf{x},t)\hat{T}^{-1} = \eta_{t_\pi}\hat{\phi}_{1,3}(\mathbf{x},-t) \qquad ; \ \hat{T}\hat{\phi}_2(\mathbf{x},t)\hat{T}^{-1} = -\eta_{t_\pi}\hat{\phi}_2(\mathbf{x},-t)$$

The additional minus sign if $i = 2$ is needed in the charge conjugation since charge conjugation takes $\hat{\phi} \to \hat{\phi}^{\star}$, and it is needed in the time reversal to counteract the complex conjugation of the i. Here, from Eqs. (8.13) and (8.136),

$$\eta_{p_\pi} = -1 \qquad ; \ \eta_{c_\pi} = +1 \qquad ; \ \eta_{t_\pi} = -1$$

The last relation is chosen to preserve CPT invariance.[9]

Let us first verify the parity invariance of $\hat{\mathcal{L}}_I(x)$. The behavior of the nucleon field $\hat{\psi} = \begin{pmatrix} \hat{\psi}_p \\ \hat{\psi}_n \end{pmatrix}$ under $\hat{P}$ is given in Eqs. (8.62)

$$\hat{P}\hat{\psi}(\mathbf{x},t)\hat{P}^{-1} = i\eta_p\gamma_4\hat{\psi}(-\mathbf{x},t)$$
$$\hat{P}\hat{\bar{\psi}}(\mathbf{x},t)\hat{P}^{-1} = -i\eta_p^*\hat{\bar{\psi}}(-\mathbf{x},t)\gamma_4$$

Then, with $|\eta_p|^2 = 1$,

$$\hat{P}\hat{\mathcal{L}}_I(x_\mu)\hat{P}^{-1} = \eta_{p_\pi}\left[ig\hat{\bar{\psi}}\,\gamma_4\gamma_5\gamma_4\boldsymbol{\tau}\,\hat{\psi}\cdot\hat{\phi} + if\hat{\bar{\psi}}\,\gamma_4\gamma_5\gamma_\mu\gamma_4\boldsymbol{\tau}\,\hat{\psi}\cdot\frac{\partial}{\partial x_\mu}\hat{\phi}(-\mathbf{x},t) \right]$$

---

[8]Note $\boldsymbol{\tau}\cdot\hat{\phi} = \sqrt{2}\,(\tau_-\hat{\phi}^{\star} + \tau_+\hat{\phi}) + \tau_3\hat{\phi}_3$, where

$$\tau_- = \frac{1}{2}(\tau_1 - i\tau_2) = \begin{pmatrix} 0 & 0 \\ 1 & 0 \end{pmatrix} \qquad ; \ \tau_+ = \frac{1}{2}(\tau_1 + i\tau_2) = \begin{pmatrix} 0 & 1 \\ 0 & 0 \end{pmatrix}$$

[9]In the interaction picture, $\partial\hat{\phi}^{\star}/\partial t$ is given by the explicit time derivative of $\hat{\phi}^{\star}(\mathbf{x},t)$, and it transforms according to Eqs. (8.47); there are analogous relations for $\partial\hat{\phi}_3/\partial t$.

Now use $\{\gamma_4, \gamma_5\} = 0$, $\{\boldsymbol{\gamma}, \gamma_5\} = 0$, and $\partial/\partial x_i = -\partial/\partial(-x_i)$ to obtain

$$\hat{P}\hat{\mathcal{L}}_I(x_\mu)\hat{P}^{-1} = \hat{\mathcal{L}}_I(x'_\mu) \qquad ; \; x'_\mu = (-\mathbf{x}, ict)$$

For charge conjugation, we use the behavior of the nucleon field in Eqs. (8.66)

$$\hat{C}\hat{\psi}_\alpha(x)\hat{C}^{-1} = \eta_c^\star[\hat{\bar{\psi}}(x)\gamma_4\gamma_2]_\alpha$$
$$\hat{C}\hat{\bar{\psi}}_\alpha(x)\hat{C}^{-1} = \eta_c[\gamma_4\gamma_2\hat{\psi}(x)]_\alpha$$

We assume the current is normal-ordered as in Eqs. (8.80)–(8.85), which gets rid of any c-number left over from the anticommutator. Then, with $|\eta_c|^2 = 1$,

$$\hat{C}\hat{\mathcal{L}}_I(x_\mu)\hat{C}^{-1} = ig\hat{\bar{\psi}}\,S_c^{-1}\gamma_5^T S_c\,\boldsymbol{\tau}^T\,\hat{\psi}\cdot\hat{C}\,\hat{\boldsymbol{\phi}}\,\hat{C}^{-1}$$
$$+\, if\hat{\bar{\psi}}\,S_c^{-1}(\gamma_5\gamma_\mu)^T S_c\,\boldsymbol{\tau}^T\,\hat{\psi}\cdot\frac{\partial}{\partial x_\mu}\hat{C}\,\hat{\boldsymbol{\phi}}\,\hat{C}^{-1}$$

From Eqs. (8.68)–(8.70), $S_c = \gamma_2\gamma_4$ and $S_c^{-1} = \gamma_4\gamma_2$ are the charge-conjugation matrices with the properties

$$S_c^{-1}\gamma_5^T S_c = S_c^{-1}\gamma_5 S_c = \gamma_5$$
$$S_c^{-1}(\gamma_5\gamma_\mu)^T S_c = S_c^{-1}\gamma_\mu^T\gamma_5^T S_c = S_c^{-1}\gamma_\mu^T S_c\gamma_5 = -\gamma_\mu\gamma_5 = \gamma_5\gamma_\mu$$

Now use for the Pauli matrices $\boldsymbol{\tau}^T = (\tau_1, -\tau_2, \tau_3)$, and we obtain

$$\hat{C}\hat{\mathcal{L}}_I(x)\hat{C}^{-1} = \hat{\mathcal{L}}_I(x)$$

For time reversal, we use $\hat{T}i\hat{T}^{-1} = -i$, and Eqs. (8.76)

$$\hat{T}\hat{\psi}(\mathbf{x}, t)\hat{T}^{-1} = i\eta_t\gamma_1\gamma_3\hat{\psi}(\mathbf{x}, -t)$$
$$\hat{T}\hat{\bar{\psi}}(\mathbf{x}, t)\hat{T}^{-1} = -i\eta_t^\star\hat{\bar{\psi}}(\mathbf{x}, -t)\gamma_3\gamma_1$$

Then, with $|\eta_t|^2 = 1$, we obtain

$$\hat{T}\hat{\mathcal{L}}_I(x_\mu)\hat{T}^{-1} = -ig\hat{\bar{\psi}}\,S_t^{-1}\gamma_5^\star S_t\,\boldsymbol{\tau}^\star\,\hat{\psi}\cdot\hat{T}\,\hat{\boldsymbol{\phi}}\,\hat{T}^{-1}$$
$$-\,if\hat{\bar{\psi}}\,S_t^{-1}(\gamma_5\gamma_\mu)^\star S_t\,\boldsymbol{\tau}^\star\,\hat{\psi}\cdot\hat{T}\frac{\partial}{\partial x_\mu}\hat{\boldsymbol{\phi}}\,\hat{T}^{-1}$$

From Eqs. (8.77)–(8.79), $S_t = \gamma_1\gamma_3$ and $S_t^{-1} = \gamma_3\gamma_1$ are the time-reversal matrices with the properties

$$S_t^{-1}\gamma_5^\star S_t = S_t^{-1}\gamma_5 S_t = \gamma_5$$
$$S_t^{-1}(\gamma_5\gamma_\mu)^\star S_t = S_t^{-1}\gamma_5\gamma_\mu^\star S_t = \gamma_5 S_t^{-1}\gamma_\mu^\star S_t = \gamma_5\gamma_\mu$$

Now use for the Pauli matrices $\boldsymbol{\tau}^* = (\tau_1, -\tau_2, \tau_3)$, and we obtain[10]

$$\hat{T}\hat{\mathcal{L}}_I(x_\mu)\hat{T}^{-1} = \hat{\mathcal{L}}_I(x'_\mu) \qquad ; \; x'_\mu = (\mathbf{x}, -ict)$$

This establishes the fact that $\hat{\mathcal{L}}_I(x)$ is *separately* invariant under $\hat{C}$ and $\hat{T}$.

Problem 8.6 Start from the assumed transformation properties of the interaction hamiltonian density in Eqs. (8.131), and derive the implied symmetry properties of the scattering operator in Eqs. (8.133).

Solution to Problem 8.6

It is assumed that the interaction hamiltonian density in the interaction picture has the transformation properties indicated in Eqs. (8.131)

$$\hat{C}\hat{\mathcal{H}}_1(x)\hat{C}^{-1} = \hat{\mathcal{H}}_1(x)$$
$$\hat{P}\hat{\mathcal{H}}_1(\mathbf{x}, t)\hat{P}^{-1} = \hat{\mathcal{H}}_1(-\mathbf{x}, t)$$
$$\hat{T}\hat{\mathcal{H}}_1(\mathbf{x}, t)\hat{T}^{-1} = \hat{\mathcal{H}}_1(\mathbf{x}, -t)$$

The scattering operator is given in terms of $\hat{\mathcal{H}}_1(x)$ in Eq. (8.132)

$$\hat{S} = \sum_{n=0}^{\infty} \frac{1}{n!} \left(\frac{-i}{\hbar c}\right)^n \int \cdots \int d^4x_1 \cdots d^4x_n P[\hat{\mathcal{H}}(x_1)\hat{\mathcal{H}}(x_2) \cdots \hat{\mathcal{H}}(x_n)]$$

where $d^4x = d^3xcdt$ and $P[\cdots]$ indicates the time-ordered P-product of the operators. Consider what the discrete symmetries do to $\hat{S}$:

- Since $\hat{C}$ leaves $\hat{\mathcal{H}}_1(x)$ invariant, it also leaves $\hat{S}$ invariant

$$\hat{C}\hat{S}\hat{C}^{-1} = \hat{S}$$

- Parity reflects the spatial coordinate; however, a change in dummy integration variable $\mathbf{x}_i \to -\mathbf{x}_i$, with $\int d^3x_i \to \int d^3x_i$, restores $\hat{S}$ to its initial form.[11] Hence $\hat{P}$ also leaves $\hat{S}$ invariant

$$\hat{P}\hat{S}\hat{P}^{-1} = \hat{S}$$

- Time reversal is a little more subtle. A change of dummy integration variable $t_i \to -t_i$ restores the time integrals to their original form; however, the operators now appear in an *anti-time-ordered* form, with the latest operator now appearing on the right. We indicate this with

---

[10]Recall $\eta_{t_\pi} = -1$, and remember the complex conjugation of $x_\mu = (\mathbf{x}, ict)$.

[11]Note that under the reflection of a component $\int_{-\infty}^{\infty} dx_i \to -\int_{\infty}^{-\infty} dx_i = \int_{-\infty}^{\infty} dx_i$.

the ordering operation $\tilde{P}[\cdots]$. Furthermore, since $\hat{T}$ is anti-unitary, one has

$$\hat{T}(-i)^n \hat{T}^{-1} = i^n$$

Thus

$$\hat{T}\hat{S}\hat{T}^{-1} = \sum_{n=0}^{\infty} \frac{1}{n!} \left(\frac{i}{\hbar c}\right)^n \int \cdots \int d^4 x_1 \cdots d^4 x_n$$
$$\times \tilde{P}[\hat{\mathcal{H}}(x_1)\hat{\mathcal{H}}(x_2)\cdots \hat{\mathcal{H}}(x_n)]$$

This is the *adjoint* of the scattering operator. Therefore

$$\hat{T}\hat{S}\hat{T}^{-1} = \hat{S}^\dagger$$

These are the stated results in Eqs. (8.133).

Problem 8.7 Consider the reaction $a + b \rightarrow c + d$ for spinless particles, with S-matrix element $\langle \mathbf{k}_c \, \mathbf{k}_d | \hat{S} | \mathbf{k}_a \, \mathbf{k}_b \rangle$.

(a) Show the time-reversed states are given by

$$\hat{T}|\mathbf{k}_a \, \mathbf{k}_b\rangle = \eta_{t_a}\eta_{t_b}| - \mathbf{k}_a, -\mathbf{k}_b\rangle \qquad ; \ \hat{T}|\mathbf{k}_c \, \mathbf{k}_d\rangle = \eta_{t_c}\eta_{t_d}| - \mathbf{k}_c, -\mathbf{k}_d\rangle$$

(b) Show that time-reversal symmetry of the scattering operator in the last of Eqs. (8.133) implies the following condition on the S-matrix elements

$$\langle \mathbf{k}_c \, \mathbf{k}_d | \hat{S} | \mathbf{k}_a \, \mathbf{k}_b \rangle = \eta_{t_a}^* \eta_{t_b}^* \eta_{t_c} \eta_{t_d} \langle -\mathbf{k}_a, -\mathbf{k}_b | \hat{S} | - \mathbf{k}_c, -\mathbf{k}_d \rangle$$

Solution to Problem 8.7

(a) Equations (8.42) give the behavior of the creation operators under time-reversal

$$\hat{T}a_{\mathbf{k}}^\dagger \hat{T}^{-1} = \eta_t a_{-\mathbf{k}}^\dagger$$

It follows that the effect of $\hat{T}$ on the two-particle state is given by

$$\hat{T}|\mathbf{k}_a \, \mathbf{k}_b\rangle = \hat{T}a_{\mathbf{k}_a}^\dagger a_{\mathbf{k}_b}^\dagger |0\rangle$$
$$= \hat{T}a_{\mathbf{k}_a}^\dagger \hat{T}^{-1}\hat{T}a_{\mathbf{k}_b}^\dagger \hat{T}^{-1}\hat{T}|0\rangle$$
$$= \eta_{t_a}\eta_{t_b}a_{-\mathbf{k}_a}^\dagger a_{-\mathbf{k}_b}^\dagger |0\rangle$$
$$= \eta_{t_a}\eta_{t_b}| - \mathbf{k}_a, -\mathbf{k}_b\rangle$$

In a similar manner

$$\hat{T}|\mathbf{k}_c\,\mathbf{k}_d\rangle = \eta_{t_c}\eta_{t_d}|-\mathbf{k}_c, -\mathbf{k}_d\rangle$$

(b) Insert $\hat{T}^{-1}\hat{T}$ in the S-matrix element

$$\langle\mathbf{k}_c\,\mathbf{k}_d|\hat{S}|\mathbf{k}_a\,\mathbf{k}_b\rangle = \langle\mathbf{k}_c\,\mathbf{k}_d|\hat{T}^{-1}\hat{T}\hat{S}\hat{T}^{-1}\hat{T}|\mathbf{k}_a\,\mathbf{k}_b\rangle$$

Use the anti-unitarity of $\hat{T}$, and the properties of the adjoint,

$$\langle\mathbf{k}_c\,\mathbf{k}_d|\hat{T}^{-1}\,\hat{T}\hat{S}\hat{T}^{-1}\,\hat{T}|\mathbf{k}_a\,\mathbf{k}_b\rangle = \langle\mathbf{k}_c\,\mathbf{k}_d|\hat{T}^\dagger\,\hat{T}\hat{S}\hat{T}^{-1}\,\hat{T}|\mathbf{k}_a\,\mathbf{k}_b\rangle^\star$$
$$= \langle T\mathbf{k}_c\,\mathbf{k}_d|\hat{T}\hat{S}\hat{T}^{-1}\,\hat{T}|\mathbf{k}_a\,\mathbf{k}_b\rangle^\star$$

Now use the behavior of the scattering operator under time-reversal

$$\hat{T}\hat{S}\hat{T}^{-1} = \hat{S}^\dagger$$

With another use of the properties of the adjoint, the result is

$$\langle\mathbf{k}_c\,\mathbf{k}_d|\hat{S}|\mathbf{k}_a\,\mathbf{k}_b\rangle = \langle T\mathbf{k}_c\,\mathbf{k}_d|\hat{S}^\dagger\,\hat{T}|\mathbf{k}_a\,\mathbf{k}_b\rangle^\star = \langle T\mathbf{k}_a\,\mathbf{k}_b|\hat{S}\,\hat{T}|\mathbf{k}_c\,\mathbf{k}_d\rangle$$

Part (a) gives the effect of the time-reversal operator on the states, where the last matrix element is calculated from $\hat{T}|\mathbf{k}_a\,\mathbf{k}_b\rangle$, and the above implies

$$\langle\mathbf{k}_c\,\mathbf{k}_d|\hat{S}|\mathbf{k}_a\,\mathbf{k}_b\rangle = \eta_{t_a}^\star\eta_{t_b}^\star\eta_{t_c}\eta_{t_d}\langle-\mathbf{k}_a, -\mathbf{k}_b|\hat{S}|-\mathbf{k}_c, -\mathbf{k}_d\rangle$$

which is the desired result. Note the appropriate reversal of the initial and final states in this relation.

Problem 8.8 Show the two results on the right in Eqs. (8.42) follow from those on the left, even though $\hat{T}$ is anti-unitary.

Solution to Problem 8.8

Consider the relation

$$\hat{T}a_{\mathbf{k}}^\dagger\hat{T}^{-1} = \eta_t a_{-\mathbf{k}}^\dagger$$

Take an arbitrary matrix element, and use the definition of the adjoint on the r.h.s.

$$\langle b|\hat{T}a_{\mathbf{k}}^\dagger\hat{T}^{-1}|a\rangle = \langle b|\eta_t a_{-\mathbf{k}}^\dagger|a\rangle$$
$$= \langle a_{-\mathbf{k}}b|\eta_t|a\rangle = \langle a|\eta_t^\star a_{-\mathbf{k}}|b\rangle^\star$$

Use the definition of the adjoint (twice) on the l.h.s.

$$\langle b|\hat{T}a_{\mathbf{k}}^\dagger\hat{T}^{-1}|a\rangle = \langle a_{\mathbf{k}}T^\dagger b|\hat{T}^{-1}|a\rangle$$

Now use the anti-unitarity of $\hat{T}$, and then the properties of the adjoint again

$$\langle a_{\mathbf{k}} T^\dagger b | \hat{T}^{-1} | a \rangle = \langle a_{\mathbf{k}} T^\dagger b | \hat{T}^\dagger | a \rangle^\star$$
$$= \langle T a_{\mathbf{k}} T^\dagger b | a \rangle^\star = \langle a | \hat{T} a_{\mathbf{k}} \hat{T}^\dagger | b \rangle$$

Use the anti-unitarity of $\hat{T}$ again

$$\langle a | \hat{T} a_{\mathbf{k}} \hat{T}^\dagger | b \rangle = \langle a | \hat{T} a_{\mathbf{k}} \hat{T}^{-1} | b \rangle^\star$$

Hence, for the l.h.s.

$$\langle b | \hat{T} a_{\mathbf{k}}^\dagger \hat{T}^{-1} | a \rangle = \langle a | \hat{T} a_{\mathbf{k}} \hat{T}^{-1} | b \rangle^\star$$

We have thus shown that

$$\langle a | \hat{T} a_{\mathbf{k}} \hat{T}^{-1} | b \rangle^\star = \langle a | \eta_t^* a_{-\mathbf{k}} | b \rangle^\star$$

From this, we can conclude that

$$\hat{T} a_{\mathbf{k}} \hat{T}^{-1} = \eta_t^* a_{-\mathbf{k}}$$

which is the desired result.

Problem 8.9 (a) Construct the interaction hamiltonian density $\mathcal{H}_W(x)$ from the interaction lagrangian density $\mathcal{L}_W(x)$ in Eq. (8.164), and show that it is a scalar under the combined operations TCP as in Eq. (8.175);[12]
(b) Repeat for $\mathcal{H}_S(x)$. The free lagrangian density for the charged scalar field is $\mathcal{L}_0/c^2 = -(\partial\phi^*/\partial x_\mu)(\partial\phi/\partial x_\mu) - (mc/\hbar)^2\phi^*\phi$. Note Eqs. (8.47).

Solution to Problem 8.9

(a) The interaction lagrangian density $\mathcal{L}_W(x)$ in Eqs. (8.164) is[13]

$$\mathcal{L}_W(x) = \left[\bar{\psi}_1(\bar{a} + i\bar{b}\gamma_5)\psi_2\right]\left[\bar{\psi}_3(\bar{c} + i\bar{d}\gamma_5)\psi_4\right]$$
$$+ \left[\bar{\psi}_1 i(\bar{e} + \bar{f}\gamma_5)\gamma_\mu\psi_2\right]\left[\bar{\psi}_3 i(\bar{g} + \bar{h}\gamma_5)\gamma_\mu\psi_4\right]$$
$$+ \bar{k}\left[\bar{\psi}_1\sigma_{\mu\nu}\psi_2\right]\left[\bar{\psi}_3\sigma_{\mu\nu}\psi_4\right] + \text{h.a.}$$

Since there are no derivative couplings, the canonical momentum density here comes from the free lagrangian density, which with four types of Dirac

---

[12] Recall we are henceforth suppressing the carets on the operators in abstract Hilbert space.

[13] Here we denote the coupling constants in the lagrangian density by $\{\bar{a}, \bar{b}, \bar{c}, \cdots, \bar{k}\}$ to avoid confusion with the speed of light.

particles is given by[14]

$$\mathcal{L}_0(x) = -\hbar c \sum_{i=1}^{4} \bar{\psi}_i \left(\gamma_\mu \frac{\partial}{\partial x_\mu} + M_i \right) \psi_i$$

The canonical momentum densities are then

$$\Pi_i = \frac{\partial \mathcal{L}}{\partial(\partial \psi_i / \partial t)} = i\hbar \psi_i^\dagger$$

The hamiltonian density follows as

$$\mathcal{H}_W(x) = \sum_{i=1}^{4} \Pi_i \frac{\partial \psi_i}{\partial t} - [\mathcal{L}_0(x) + \mathcal{L}_W(x)]$$
$$= \mathcal{H}_0(x) - \mathcal{L}_W(x)$$

where $\mathcal{H}_0(x)$ is the free Dirac hamiltonian density given by

$$\mathcal{H}_0(x) = \sum_{i=1}^{4} \psi_i^\dagger \left(c\boldsymbol{\alpha} \cdot \mathbf{p} + \beta m_i c^2 \right) \psi_i$$

with $\mathbf{p} = (\hbar/i)\boldsymbol{\nabla}$. Since $\mathcal{H}_0(x)$ has the appropriate invariance *separately* under P, C, and T, it follows that the hamiltonian density $\mathcal{H}_W(x)$ is a scalar under the combined operations TCP as in Eq. (8.175).

(b) The interaction lagrangian $\mathcal{L}_S(x)$ in Eqs. (8.164) is given by

$$\mathcal{L}_S(x) = \bar{\psi}_1(\bar{a} + i\bar{b}\gamma_5)\psi_2 \phi + \bar{\psi}_1 i(\bar{c} + \bar{d}\gamma_5)\gamma_\mu \psi_2 \frac{\partial \phi}{\partial x_\mu} + \text{h.a.}$$

Since there is now a derivative coupling, we have to exercise some care in passing to the hamiltonian density. The free lagrangian density for the charged scalar field is

$$\mathcal{L}_0^\phi = -c^2 \left[\left(\frac{\partial \phi^\star}{\partial x_\mu} \right) \left(\frac{\partial \phi}{\partial x_\mu} \right) + m_s^2 \phi^\star \phi \right]$$

While the Dirac canonical momentum densities are unchanged from part (a), the canonical momentum densities for the charged scalar field are now obtained as

$$\Pi_\phi = \frac{\partial \mathcal{L}}{\partial(\partial \phi / \partial t)} = \frac{\partial \phi^\star}{\partial t} + \frac{1}{ic} \bar{\psi}_1 i(\bar{c} + \bar{d}\gamma_5)\gamma_4 \psi_2$$
$$\Pi_{\phi^\star} = \frac{\partial \mathcal{L}}{\partial(\partial \phi^\star / \partial t)} = \frac{\partial \phi}{\partial t} + \frac{1}{ic} \bar{\psi}_2 i(\bar{c}^\star + \bar{d}^\star \gamma_5)\gamma_4 \psi_1$$

---

[14]See Eq. (5.92) in [Walecka (2010)].

Notice that

$$\Pi_{\phi^\star} = \Pi_\phi^\star$$

The hamiltonian density is then given by

$$\mathcal{H}_S(x) = \sum_{i=1}^{4} \Pi_{\psi_i} \frac{\partial \psi_i}{\partial t} + \Pi_\phi \frac{\partial \phi}{\partial t} + \Pi_{\phi^\star} \frac{\partial \phi^\star}{\partial t} - [\mathcal{L}_0(x) + \mathcal{L}_S(x)]$$

This is evaluated as

$$\mathcal{H}_S(x) = \mathcal{H}_0(x) - \mathcal{L}_S(x) + \bar{\psi}_1 i(\bar{c} + \bar{d}\gamma_5)\gamma_4\psi_2 \frac{\partial \phi}{\partial x_4} + \bar{\psi}_2 i(\bar{c}^\star + \bar{d}^\star\gamma_5)\gamma_4\psi_1 \frac{\partial \phi^\star}{\partial x_4}$$

where

$$\mathcal{H}_0(x) = \sum_{i=1}^{4} \psi_i^\dagger \left(c\boldsymbol{\alpha} \cdot \mathbf{p} + \beta m_i c^2 \right) \psi_i + \left[\frac{\partial \phi^\star}{\partial t} \frac{\partial \phi}{\partial t} + c^2 \boldsymbol{\nabla}\phi^\star \cdot \boldsymbol{\nabla}\phi + m_s^2 c^2 \, \phi^\star\phi \right]$$

The symmetry properties of $\mathcal{H}_S(x)$ in the interaction picture are now analyzed as follows:[15]

- The free hamiltonian density $\mathcal{H}_0(x)$ has the appropriate invariance *separately* under P, C, and T;
- As we have seen, the strong lagrangian density $\mathcal{L}_S(x)$ is a scalar under the combined operations TCP;
- The last two terms in $\mathcal{H}_S(x)$ transform in the same manner as the terms involving the fourth-component of the derivative in $\mathcal{L}_S(x)$, and, as shown in the text in Eqs. (8.173), these yield a scalar under TCP.

Hence, in *summary*, the hamiltonian density $\mathcal{H}_S(x)$ in the interaction picture is a scalar under the combined operations TCP.

---

[15]In the interaction picture, $\partial\phi^\star/\partial t$ is given by Eq. (8.46)

$$\frac{\partial \phi^\star(\mathbf{x}, t)}{\partial t} = \frac{i}{\hbar}[H_0, \, \phi^\star(\mathbf{x}, t)] = \sum_{\mathbf{k}} \left(\frac{\hbar\omega_k}{2\Omega} \right)^{1/2} i \left(a_{\mathbf{k}}^\dagger e^{-ik\cdot x} - b_{\mathbf{k}} e^{ik\cdot x} \right)$$

This is the quantum analog of the classical field expansion, and provides a representation of the canonical equal-time commutation relation

$$[\phi(\mathbf{x}, t), \, \Pi_\phi(\mathbf{x}', t')]_{t=t'} = i\hbar\,\delta^{(3)}(\mathbf{x} - \mathbf{x}')$$

(Compare the discussion in Sec. 8.3 of Vol. II.) The quantity $\partial\phi^\star(\mathbf{x}, t)/\partial t$ is then the explicit time derivative of $\phi^\star(\mathbf{x}, t)$ and transforms according to Eqs. (8.47). Note that, in fact, the term containing time derivatives *disappears* from the interaction part of $\mathcal{H}_S(x)$.

Chapter 9

Heisenberg Picture

Problem 9.1 Start from Eq. (9.7) and show

$$TU(0, -\infty)T^{-1} = U(+\infty, 0)^\dagger = U(0, +\infty)$$

Solution to Problem 9.1

The time-development operator for $t > t_0$ is given in Eq. (9.7) as

$$U(t, t_0) = \sum_{n=0}^{\infty} \frac{1}{n!} \left(\frac{-i}{\hbar c} \right)^n \int_{t_0}^{t} \cdots \int_{t_0}^{t} d^4 x_1 \cdots d^4 x_n P\left[\mathcal{H}_I(x_1) \cdots \mathcal{H}_I(x_n) \right]$$

Here $\mathcal{H}_I(x)$ is the interaction hamiltonian in the interaction representation, and P is the time-ordering operation, which orders the operators so that the operator with the latest time appears on the left.

The behavior of $\mathcal{H}_I(x)$ under time reversal is given by Eqs. (8.131) as

$$T\mathcal{H}_1(\mathbf{x}, t)T^{-1} = \mathcal{H}_1(\mathbf{x}, -t)$$

We recall that T is an anti-unitary operator satisfying

$$TiT^{-1} = -i$$

Consider $TU(0, -\infty)T^{-1}$

$$TU(0, -\infty)T^{-1} = \sum_{n=0}^{\infty} \frac{1}{n!} \left(\frac{i}{\hbar c} \right)^n \int_{-\infty}^{0} \cdots \int_{-\infty}^{0} d^4 x_1 \cdots d^4 x_n$$
$$\times \tilde{P}\left[\mathcal{H}_I(\mathbf{x}_1, -t_1) \cdots \mathcal{H}_I(\mathbf{x}_n, -t_n) \right]$$

Here $\tilde{P}$ is the anti-time-ordered product which orders the operators with the *earliest* times to the left. Now change dummy variables with $t_i \to -t_i$, and

$$\int_{-\infty}^{0} dt_i \to \int_{0}^{\infty} dt_i$$

It follows that

$$TU(0, -\infty)T^{-1} = \sum_{n=0}^{\infty} \frac{1}{n!} \left(\frac{i}{\hbar c}\right)^n \int_{0}^{\infty} \cdots \int_{0}^{\infty} d^4x_1 \cdots d^4x_n$$
$$\times \tilde{P}\left[\mathcal{H}_I(x_1) \cdots \mathcal{H}_I(x_n)\right]$$

This is the *adjoint* of the expression for $U(+\infty, 0)$, and therefore

$$TU(0, -\infty)T^{-1} = U(+\infty, 0)^{\dagger}$$

If $t < t_0$, one can re-write Eq. (6.184) as

$$c_n^{(i)}(t) = \delta_{ni} + \frac{i}{\hbar} \sum_m \int_{t}^{t_0} dt_1 \, \langle n|\hat{H}_I(t_1)|m\rangle c_m^{(i)}(t_1)$$

Now $c_n^{(i)}(t)$ satisfies the same differential equation in t; however, t appears as the *lower* limit in the integral. A repetition of the arguments in section 6.3 then leads to the following expression for $U(0, +\infty)$ [1]

$$U(0, +\infty) = \sum_{n=0}^{\infty} \frac{1}{n!} \left(\frac{i}{\hbar c}\right)^n \int_{0}^{\infty} \cdots \int_{0}^{\infty} d^4x_1 \cdots d^4x_n \tilde{P}\left[\mathcal{H}_I(x_1) \cdots \mathcal{H}_I(x_n)\right]$$

A comparison with the above gives

$$U(0, +\infty) = U(+\infty, 0)^{\dagger}$$

This illustrates the general property of the time-development operator, which follows directly from Eq. (9.6), that

$$U(t, t_0) = U(t_0, t)^{\dagger}$$

Therefore, in summary,

$$TU(0, -\infty)T^{-1} = U(+\infty, 0)^{\dagger} = U(0, +\infty)$$

---

[1] Note the $\tilde{P}$; readers should go through these arguments (see chapter 4 in Vol. II).

Problem 9.2 (a) Explicitly demonstrate the result in Eqs. (9.22)–(9.23), used to prove Theorem Ia, up through $\nu = 4$;

(b) Demonstrate the corresponding result used to prove Theorem Ib.

Solution to Problem 9.2

(a) The goal is to show explicitly up through $\nu = 4$ that the νth term in the series in Eq. (9.23)

$$I_\nu(t) \equiv \frac{1}{\nu!}\left(\frac{-i}{\hbar c}\right)^\nu \int_{-\infty}^{\infty}\cdots\int_{-\infty}^{\infty} d^4x_1\cdots d^4x_\nu P\left[\mathcal{H}_I(x_1)\cdots\mathcal{H}_I(x_\nu)O_I(t)\right]$$

can be written as those terms in the double sum in Eq. (9.22) with $n+m = \nu$

$$
\begin{aligned}
I_\nu(t) = {}&\left(\frac{-i}{\hbar c}\right)^\nu \sum_{n=0}^{\infty}\sum_{m=0}^{\infty}\delta_{\nu,m+n}\frac{1}{n!}\frac{1}{m!}\\
&\times \int_{t}^{\infty}\cdots\int_{t}^{\infty} d^4x_1\cdots d^4x_n P\left[\mathcal{H}_I(x_1)\cdots\mathcal{H}_I(x_n)\right]O_I(t)\\
&\times \int_{-\infty}^{t}\cdots\int_{-\infty}^{t} d^4y_1\cdots d^4y_m P\left[\mathcal{H}_I(y_1)\cdots\mathcal{H}_I(y_m)\right]
\end{aligned}
$$

We analyze the most complex term $I_4(t)$ to show how the proof goes. $I_4(t)$ is a four-dimensional integral in the time that receives additive contributions from the various time regions:

- There is one contribution where all $t_i > t$. In this contribution, all the operators $\mathcal{H}_I(x_i)$ stand to the left of $O_I(t)$, and they are themselves time-ordered. This contribution is provided by $[n, m] = [4, 0]$;
- Similarly, there is one contribution where all $t_i < t$. This is provided by $[n, m] = [0, 4]$;
- There are $4!/3!1! = 4$ contributions with one $t_i > t$ and the other three $t_i < t$. The operators $\mathcal{H}_I(x_i)$ corresponding to the latter three times stand to the right of $O_I(t)$, and they are themselves time-ordered. All four of these contributions are identical by a re-labeling of dummy integration variables. This contribution is provided by $[n, m] = [1, 3]$;
- Similarly, there are $4!/3!1! = 4$ identical contributions with one $t_i < t$ and the other three $t_i > t$. This is given by $[n, m] = [3, 1]$;
- There are $4!/2!2! = 6$ identical contributions with two times $t_i > t$ and two times $t_i < t$. This contribution is provided by $[n, m] = [2, 2]$.

This exhausts the possibilities for $n + m = 4$ and provides a complete accounting of all the time regions in the multiple integral.

The terms with $\nu = 0, 1, 2, 3$ are analyzed in a similar fashion.

(b) The extension involves establishing the equality of the expression

$$I_\nu(t_1, t_2) \equiv \frac{1}{\nu!} \left(\frac{-i}{\hbar c}\right)^\nu \int_{-\infty}^\infty \cdots \int_{-\infty}^\infty d^4x_1 \cdots d^4x_\nu$$
$$\times P\left[\mathcal{H}_I(x_1) \cdots \mathcal{H}_I(x_\nu) O_I(t_1) O_I(t_2)\right]$$

and its decomposition into[2]

$$I_\nu(t_1, t_2) = \left(\frac{-i}{\hbar c}\right)^\nu \sum_{n=0}^\infty \sum_{p=0}^\infty \sum_{m=0}^\infty \delta_{\nu, m+p+n} \frac{1}{n!}\frac{1}{p!}\frac{1}{m!}$$

$$\times \int_{t_1}^\infty \cdots \int_{t_1}^\infty d^4x_1 \cdots d^4x_n P\left[\mathcal{H}_I(x_1) \cdots \mathcal{H}_I(x_n)\right] O_I(t_1)$$

$$\times \int_{t_2}^{t_1} \cdots \int_{t_2}^{t_1} d^4y_1 \cdots d^4y_p P\left[\mathcal{H}_I(y_1) \cdots \mathcal{H}_I(y_p)\right] O_I(t_2)$$

$$\times \int_{-\infty}^{t_2} \cdots \int_{-\infty}^{t_2} d^4z_1 \cdots d^4z_m P\left[\mathcal{H}_I(z_1) \cdots \mathcal{H}_I(z_m)\right]$$

The proof proceeds exactly as in part (a), with a slightly more complicated partition of ν into $[n, p, m]$. We will not do all the terms in $I_4(t_1, t_2)$, but simply consider a couple of them to see how it goes:

- There is one contribution where all $t_i > t_1$. In this contribution, all the operators $\mathcal{H}_I(x_i)$ stand to the left of $O_I(t_1)$, and they are themselves time-ordered. This contribution is provided by $[n, p, m] = [4, 0, 0]$;
- Similarly, there is one contribution where all $t_i < t_2$. This is provided by $[n, p, m] = [0, 0, 4]$;
- There are $4!/1!3!1! = 4$ contributions with one $t_i > t_1$ and the other three with $t_2 < t_i < t_1$. The operators $\mathcal{H}_I(x_i)$ corresponding to the latter three times stand to the right of $O_I(t_1)$ and to the left of $O_I(t_2)$, and they are themselves time-ordered. All four of these contributions are identical by a re-labeling of dummy integration variables. This contribution is provided by $[n, p, m] = [1, 3, 0]$;
- Similarly, there are $4!/1!3!1! = 4$ identical contributions with one $t_i < t_2$ and the other three satisfying $t_1 > t_i > t_2$. This is given by $[n, p, m] = [0, 3, 1]$.

The proof for $I_4(t_1, t_2)$ proceeds through a complete enumeration of the fifteen partitions of ν into $[n, p, m]$.

---

[2]We assume $t_1 > t_2$.

Problem 9.3 (a) Use the fact that the Heisenberg states $|\underline{\Psi}^{(\pm)}\rangle$ are eigenstates of the four-momentum operator to show that all other states are orthogonal to the vacuum $|\underline{0}^{(\pm)}\rangle$, assumed to be non-degenerate.[3]
(b) Repeat the argument for the stable single-particle states $|\underline{p}^{(\pm)}\rangle$.

Solution to Problem 9.3

It is shown following Eq. (9.43) that the Heisenberg states $|\underline{p}_n^{(\pm)}\rangle$ are eigenstates of the total four-momentum $P_\mu = (\mathbf{P}, iH/c)$ with eigenvalues $p_{n,\mu} = (\mathbf{p}_n, iE_n/c)$ [4]

$$P_\mu |\underline{p}_n^{(\pm)}\rangle = p_{n,\mu} |\underline{p}_n^{(\pm)}\rangle$$

Consider another state with a distinct eigenvalue

$$P_\mu |\underline{p}'_n{}^{(\pm)}\rangle = p'_{n,\mu} |\underline{p}'_n{}^{(\pm)}\rangle$$

Take the matrix element of the first equation with this state

$$\langle \underline{p}'_n{}^{(\pm)} |P_\mu| \underline{p}_n^{(\pm)}\rangle = p_{n,\mu} \langle \underline{p}'_n{}^{(\pm)} |\underline{p}_n^{(\pm)}\rangle$$

Now take the complex conjugate of the matrix element of the second equation with the first state

$$\langle \underline{p}_n^{(\pm)} |P_\mu| \underline{p}'_n{}^{(\pm)}\rangle^\star = p'^\star_{n,\mu} \langle \underline{p}_n^{(\pm)} |\underline{p}'_n{}^{(\pm)}\rangle^\star$$

Use the hermiticity of $(\mathbf{P}, H)$ to re-write this as

$$\langle \underline{p}'_n{}^{(\pm)} |P_\mu| \underline{p}_n^{(\pm)}\rangle = p'_{n,\mu} \langle \underline{p}'_n{}^{(\pm)} |\underline{p}_n^{(\pm)}\rangle$$

Subtract the previous expression for this quantity from this one to obtain

$$\left(p'_{n,\mu} - p_{n,\mu} \right) \langle \underline{p}'_n{}^{(\pm)} |\underline{p}_n^{(\pm)}\rangle = 0$$

We conclude that two eigenstates of the four-momentum operator corresponding to distinct eigenvalues are orthogonal

$$\langle \underline{p}'_n{}^{(\pm)} |\underline{p}_n^{(\pm)}\rangle = 0 \qquad ; \ p'_{n,\mu} \neq p_{n,\mu}$$

---

[3] As a nicety, you can give all particles some mass, even infinitesimal, in making this argument.

[4] The states $|\Psi\rangle$ differ from the states $|\underline{\Psi}\rangle$ only through the vacuum-vacuum phase [see Eqs. (9.30) and (9.18)].

(a) Consider the vacuum state $|\underset{\sim}{0}^{(\pm)}\rangle$ with zero four-momentum, and assume that it is non-degenerate. Every other state then has some four-momentum, and hence[5]

$$\langle \underset{\sim}{p'}_n^{(\pm)}|\underset{\sim}{0}^{(\pm)}\rangle = 0 \qquad ; \; p'_{n,\mu} \neq 0$$

(b) Assume the particle with four-momentum $p_\mu^2 = -(mc)^2$ and eigenstate $|\underset{\sim}{p}^{(\pm)}\rangle$ is stable and non-degenerate. There are then no other states with $p'^2_n = -(mc)^2$ into which it can decay. Hence

$$\langle \underset{\sim}{p'}_n^{(\pm)}|\underset{\sim}{p}^{(\pm)}\rangle = 0 \qquad ; \; p'^2_{n,\mu} \neq -(mc)^2$$

Problem 9.4 (a) Start from the expression for $U(t,t_0)$ for finite times in Eq. (9.6) and derive Eq. (9.47). Then use the iterated series in Eq. (9.7) and adiabatic damping in Eq. (9.8) to let $t_0 \to -\infty$, and derive Eq. (9.48);

(b) Verify Eq. (9.52) for $|\underset{\sim}{p}_n^{(-)}\rangle$.

Solution to Problem 9.4

(a) The time-development operator $U(t,t_0)$ is given for finite times in Eq. (9.6)

$$U(t,t_0) = e^{iH_0 t/\hbar} e^{-iH(t-t_0)/\hbar} e^{-iH_0 t_0/\hbar}$$

Translate the time by an interval τ, so that $t \to t + \tau$ and $t_0 \to t_0 + \tau$. Then

$$U(t+\tau, t_0+\tau) = e^{iH_0(t+\tau)/\hbar} e^{-iH[(t+\tau)-(t_0+\tau)]/\hbar} e^{-iH_0(t_0+\tau)/\hbar}$$
$$= e^{iH_0\tau/\hbar} U(t,t_0) e^{-iH_0\tau/\hbar}$$

This is Eq. (9.47). We will also need the group property, established in the same manner

$$U(t_1, t_2)U(t_2, t_3) = U(t_1, t_3)$$

These results imply

$$e^{iH_0\tau/\hbar} U(0,t_0) e^{-iH_0\tau/\hbar} = U(\tau, t_0+\tau) = U(\tau,0)U(0,t_0+\tau)$$

---

[5]The infrared problem in QED is handled by giving the photons an infinitesimal mass, and then, at the end, letting this mass go to zero. At each step, the orthogonality of the vacuum then holds. A more rigorous treatment of the zero-mass limit goes beyond the scope of this work.

The iterated series for $U(t, t_0)$ in Eq. (9.7) is

$$U(t, t_0) = \sum_{n=0}^{\infty} \frac{1}{n!} \left(\frac{-i}{\hbar c}\right)^n \int_{t_0}^t \cdots \int_{t_0}^t d^4 x_1 \cdots d^4 x_n P\left[\mathcal{H}_I(x_1) \cdots \mathcal{H}_I(x_n)\right]$$

It follows that

$$e^{iH_0\tau/\hbar} U(0, t_0) e^{-iH_0\tau/\hbar} = U(\tau, 0)$$

$$\times \sum_{n=0}^{\infty} \frac{1}{n!} \left(\frac{-i}{\hbar c}\right)^n \int_{t_0+\tau}^0 \cdots \int_{t_0+\tau}^0 d^4 x_1 \cdots d^4 x_n P\left[\mathcal{H}_I(x_1) \cdots \mathcal{H}_I(x_n)\right]$$

With the adiabatic damping in Eq. (9.8), one replaces

$$\mathcal{H}_I(x) \to e^{-\epsilon|t|} \mathcal{H}_I(x)$$

Now let $t_0 \to -\infty$ in the above. The adiabatic damping forces the integrand to zero at the lower limit, and hence the extra displacement by τ in that limit is irrelevant. Hence

$$e^{iH_0\tau/\hbar} U(0, -\infty) e^{-iH_0\tau/\hbar} = U(\tau, 0)$$

$$\times \sum_{n=0}^{\infty} \frac{1}{n!} \left(\frac{-i}{\hbar c}\right)^n \int_{-\infty}^0 \cdots \int_{-\infty}^0 d^4 x_1 \cdots d^4 x_n P\left[\mathcal{H}_I(x_1) \cdots \mathcal{H}_I(x_n)\right]$$

This is Eq. (9.48)

$$e^{iH_0\tau/\hbar} U(0, -\infty) e^{-iH_0\tau/\hbar} = U(\tau, 0) U(0, -\infty)$$

(b) The states $|p_n^{(\pm)}\rangle$ in Eq. (9.43) are defined through Eqs. (9.30) as

$$|p_n^{(+)}\rangle \equiv e^{-i\Phi/2} U(0, -\infty)|p_n\rangle$$

$$|p_n^{(-)}\rangle \equiv e^{+i\Phi/2} U(0, +\infty)|p_n\rangle$$

It is now clear that the only difference in the proof in Eqs. (9.44)–(9.52) is the presence of $U(0, +\infty)$ rather than $U(0, -\infty)$;[6] however, exactly the same proof as in part (a) establishes the relation

$$e^{iH_0\tau/\hbar} U(0, +\infty) e^{-iH_0\tau/\hbar} = U(\tau, 0) U(0, +\infty)$$

Therefore, exactly as in Eqs. (9.44)–(9.52),

$$H|p_n^{(-)}\rangle = E_n |p_n^{(-)}\rangle$$

---

[6] Note that $U(0, +\infty) = U(+\infty, 0)^\dagger$ (see Prob. 9.1).

Problem 9.5 Consider elastic electron scattering from the nucleon (Fig. 9.1).

Fig. 9.1   Elastic electron scattering from the nucleon.

The S-matrix for this process is given by[7]

$$S_{fi} = \frac{(2\pi)^4}{\Omega^2}\delta^{(4)}(p_1 + k_1 - p_2 - k_2)\bar{u}(\mathbf{k}_2)\gamma_\mu u(\mathbf{k}_1)\frac{4\pi\alpha}{q^2}\left(\frac{M^2}{E_1 E_2}\right)^{1/2}(\mathcal{J}_\mu)_{fi}$$

$$(\mathcal{J}_\mu)_{fi} = \left(\frac{\Omega^2 E_1 E_2}{M^2}\right)^{1/2}\langle \underset{\sim}{p}_2|J_\mu(0)|\underset{\sim}{p}_1\rangle$$

Here $e_p J_\mu(0)$ is the electromagnetic current operator for the hadronic system. The matrix element is in the Heisenberg picture; it includes all Feynman diagrams contributing to the electromagnetic structure of the nucleon.

(a) Use Lorentz invariance, current conservation, and the Dirac equation to show

$$(\mathcal{J}_\mu)_{fi} = i\bar{\mathcal{U}}(\mathbf{p}_2\lambda_2)\left[F_1(q^2)\gamma_\mu - F_2(q^2)\sigma_{\mu\nu}q_\nu + iF_3(q^2)\gamma_5\sigma_{\mu\nu}q_\nu\right]\mathcal{U}(\mathbf{p}_1\lambda_1)$$

where $\bar{\mathcal{U}}\mathcal{U} = 1$;

(b) Use hermiticity of the current to prove the form factors F_i are real $(i = 1, 2, 3)$;

(c) Use invariance under P to prove $F_3 = 0$;

(d) Use invariance under T to *separately* prove $F_3 = 0$;

(e) Make a non-relativistic reduction of the current matrix element, and give a physical interpretation of the term proportional to F_3 as an electric dipole moment.

---

[7]Compare Eq. (8.51) in Vol. II. This problem is longer, but the results form an essential part of nuclear physics.

Solution to Problem 9.5

(a) The analysis of $(\mathcal{J}_\mu)_{fi}$ proceeds as follows:

- The factor in front of the matrix element is just such as to produce the states with invariant norm in Eqs. (9.66); equivalently, it converts the Dirac spinors appearing in the matrix element to those with the invariant norm $\bar{\mathcal{U}}\mathcal{U} = 1$;
- The electromagnetic current transforms as a four-vector, thus the resulting matrix element of the current $(\mathcal{J}_\mu)_{fi}$ must be a four-vector;
- From translation invariance, the current operator has the form in Eq. (9.54)

$$J_\mu(x) = e^{-iP_\mu x_\mu/\hbar} J_\mu(0) e^{iP_\mu x_\mu/\hbar}$$

The space-time behavior of the matrix element follows as

$$\langle \underset{\sim}{p}_2 | J_\mu(x) | \underset{\sim}{p}_1 \rangle = e^{-iq_\mu x_\mu} \langle \underset{\sim}{p}_2 | J_\mu(0) | \underset{\sim}{p}_1 \rangle$$

The current is conserved

$$\frac{\partial J_\mu(x)}{\partial x_\mu} = 0$$

Evaluation of the derivative at the origin then yields the statement of current conservation in momentum space

$$q_\mu (\mathcal{J}_\mu)_{fi} = 0 \qquad ; \text{ current conservation}$$

- It is difficult to give a systematic derivation of the general form of the matrix element since so many tricks can be used in the reduction of the contribution from any Feynman diagram, no matter how complicated. The basic ideas are the following:

  - Since the particles are on the mass shell, and four-momentum is conserved

$$p_1^2 = p_2^2 = -M^2$$
$$p_2 = p_1 + q$$

there is only one independent scalar on which the form factors can depend, and this can be taken as q^2;

  - After moving $\not{p}_1$ to the right and $\not{p}_2$ to the left, they may be replaced by masses through the use of the Dirac equation[8]

$$(i\not{p}_1 + M)\mathcal{U}(\mathbf{p}_1) = 0 \qquad ; \bar{\mathcal{U}}(\mathbf{p}_2)(i\not{p}_2 + M) = 0$$

---

[8] Recall $\not{a} = \gamma_\mu a_\mu$, and here the four-momenta are $(\hbar p_2, \hbar p_1)$.

Thus the scalar forms $(\not p_1, \not p_2)$ can be eliminated from the matrix element;

- There are two independent momentum four-vectors

$$q_\mu = (p_2 - p_1)_\mu \qquad ; \ P_\mu = (p_2 + p_1)_\mu$$

 - These are to be combined with the complete set of Dirac matrices $(1, \gamma_5, \gamma_\mu, \gamma_5\gamma_\mu, \sigma_{\mu\nu})$ to produce a conserved four-vector;
 - Use the Gordon decomposition of the current, derived from $\gamma_\mu\gamma_\nu = \delta_{\mu\nu} + i\sigma_{\mu\nu}$, which states that between the Dirac spinors

$$P_\mu \doteq 2iM\gamma_\mu + i\sigma_{\mu\nu}q_\nu$$
$$\gamma_5 P_\mu \doteq i\gamma_5\sigma_{\mu\nu}q_\nu$$

 - With dedicated algebra, everything can then eventually be reduced to the three forms exhibited in the statement of the problem[9]

$$\gamma_\mu \qquad ; \ \sigma_{\mu\nu}q_\nu \qquad ; \ \gamma_5\sigma_{\mu\nu}q_\nu$$

(b) Take the complex conjugate of $(\mathcal{J}_\mu)_{fi}$, and let $q \to -q$,

$$(\mathcal{J}_\mu)_{fi}^\star = \left(\frac{\Omega^2 E_1 E_2}{M^2}\right)^{1/2} \langle \underset{\sim}{p}_2|J_\mu(0)|\underset{\sim}{p}_1\rangle^\star$$
$$= \varepsilon(\mu)\,i\bar{U}(\mathbf{p}_1\lambda_1)\left[F_1^\star(q^2)\gamma_\mu - F_2^\star(q^2)\sigma_{\mu\nu}q_\nu + iF_3^\star(q^2)\gamma_5\sigma_{\mu\nu}q_\nu\right]U(\mathbf{p}_2\lambda_2)$$
$$; \ q = p_1 - p_2$$

Here, in the second line,

$$\varepsilon(\mu) = +1 \qquad ; \ \mu = 1, 2, 3$$
$$= -1 \qquad ; \ \mu = 4$$

Now write the covariant form of $(\mathcal{J}_\mu)_{if}$

$$(\mathcal{J}_\mu)_{if} = \left(\frac{\Omega^2 E_1 E_2}{M^2}\right)^{1/2} \langle \underset{\sim}{p}_1|J_\mu(0)|\underset{\sim}{p}_2\rangle$$
$$= i\bar{U}(\mathbf{p}_1\lambda_1)\left[F_1(q^2)\gamma_\mu - F_2(q^2)\sigma_{\mu\nu}q_\nu + iF_3(q^2)\gamma_5\sigma_{\mu\nu}q_\nu\right]U(\mathbf{p}_2\lambda_2)$$
$$; \ q = p_1 - p_2$$

---

[9]While the last two forms are explicitly conserved, the matrix element of the first is also, since $\bar{U}(\mathbf{p}_2)\not q\, U(\mathbf{p}_1) = \bar{U}(\mathbf{p}_2)(\not p_2 - \not p_1)U(\mathbf{p}_1) = 0$.

These two expressions are related by the hermiticity of the current $J_\mu = (\mathbf{J}, i\rho)$

$$\langle \underset{\sim}{p}_2 | J_\mu(0) | \underset{\sim}{p}_1 \rangle^* = \varepsilon(\mu) \langle \underset{\sim}{p}_1 | J_\mu(0) | \underset{\sim}{p}_2 \rangle$$

which implies

$$(\mathscr{J}_\mu)^*_{fi} = \varepsilon(\mu) (\mathscr{J}_\mu)_{if}$$

Therefore

$$(\mathscr{J}_\mu)^*_{fi} = \varepsilon(\mu) i \bar{U}(\mathbf{p}_1 \lambda_1) \left[F_1(q^2)\gamma_\mu - F_2(q^2)\sigma_{\mu\nu}q_\nu + iF_3(q^2)\gamma_5\sigma_{\mu\nu}q_\nu \right] U(\mathbf{p}_2 \lambda_2)$$
$$; \ q = p_1 - p_2$$

Compare this result with the previous expression for this quantity

$$(\mathscr{J}_\mu)^*_{fi} = \varepsilon(\mu) i \bar{U}(\mathbf{p}_1 \lambda_1) \left[F_1^*(q^2)\gamma_\mu - F_2^*(q^2)\sigma_{\mu\nu}q_\nu + iF_3^*(q^2)\gamma_5\sigma_{\mu\nu}q_\nu \right] U(\mathbf{p}_2 \lambda_2)$$
$$; \ q = p_1 - p_2$$

Since the contributions are linearly independent, one concludes that all the form factors are *real*

$$F_i^*(q^2) = F_i(q^2) \qquad ; \ i = 1, 2, 3 \qquad ; \text{ real}$$

(c) From Eqs. (9.39) and (8.56), the effect of the parity operator on the states is given by

$$P | \underset{\sim}{p}_1, \lambda_1 \rangle = \eta_p^* | -\underset{\sim}{p}_1, -\lambda_1 \rangle$$

where $|\eta_p|^2 = 1$. Furthermore, the electromagnetic current transforms as a polar vector under spatial reflection

$$P J_\mu(0) P^{-1} = -\varepsilon(\mu) J_\mu(0)$$

Therefore, upon the insertion of $P^{-1}P$ (twice) one has

$$\langle \underset{\sim}{p}_2, \lambda_2 | J_\mu(0) | \underset{\sim}{p}_1 \lambda_1 \rangle = -\varepsilon(\mu) \langle -\underset{\sim}{p}_2, -\lambda_2 | J_\mu(0) | -\underset{\sim}{p}_1, -\lambda_1 \rangle \qquad ; \text{ parity}$$

With the insertion of the covariant form, one therefore concludes

$$(\mathscr{J}_\mu)_{fi} = -\varepsilon(\mu) i \bar{U}(-\mathbf{p}_2, -\lambda_2)$$
$$\times \left[F_1(q^2)\gamma_\mu - F_2(q^2)\sigma_{\mu\nu}q_\nu + iF_3(q^2)\gamma_5\sigma_{\mu\nu}q_\nu \right] U(-\mathbf{p}_1, -\lambda_1)$$
$$; \ q = [-\mathbf{p}_2 + \mathbf{p}_1, i(p_{20} - p_{10})]$$

Now use the following relation for the parity-reflected Dirac spinors from Eqs. (8.61)

$$\mathcal{U}(-\mathbf{p}, -\lambda) = i\gamma_4 \mathcal{U}(\mathbf{p}, \lambda)$$

This gives

$$(\mathcal{J}_\mu)_{fi} = i\bar{\mathcal{U}}(\mathbf{p}_2, \lambda_2) \left[F_1(q^2)\gamma_\mu - F_2(q^2)\sigma_{\mu\nu}q_\nu - iF_3(q^2)\gamma_5\sigma_{\mu\nu}q_\nu \right] \mathcal{U}(\mathbf{p}_1, \lambda_1)$$
$$; \ q = p_2 - p_1$$

A comparison with the initial expression

$$(\mathcal{J}_\mu)_{fi} = i\bar{\mathcal{U}}(\mathbf{p}_2\lambda_2) \left[F_1(q^2)\gamma_\mu - F_2(q^2)\sigma_{\mu\nu}q_\nu + iF_3(q^2)\gamma_5\sigma_{\mu\nu}q_\nu \right] \mathcal{U}(\mathbf{p}_1\lambda_1)$$
$$; \ q = p_2 - p_1$$

leads to the conclusion that F_3 must vanish

$$F_3(q^2) = 0 \qquad ; \ \text{parity}$$

(d) From Eqs. (9.42), (9.34), and (8.72), the effect of the time-reversal operator on the stable single-particle state is

$$T|\underset{\sim}{\mathbf{p}}_1, \lambda_1\rangle = \eta_t^* S_{\lambda_1} |-\underset{\sim}{\mathbf{p}}_1, \lambda_1\rangle$$

where $|\eta_t|^2 = 1$, and

$$S_\lambda = +1 \qquad ; \ \lambda = \downarrow$$
$$= -1 \qquad ; \ \lambda = \uparrow$$

Under time reversal, the current operator $J_\mu = (\mathbf{J}, i\rho)$ transforms as

$$TJ_\mu(0)T^{-1} = -J_\mu(0)$$

Hence with the insertion of $T^{-1}T$ (twice), and the invocation of the anti-unitarity of T, one has for the matrix element

$$\langle \underset{\sim}{\mathbf{p}}_2, \lambda_2 | J_\mu(0) | \underset{\sim}{\mathbf{p}}_1\lambda_1 \rangle = -S_{\lambda_1}S_{\lambda_2}\langle -\underset{\sim}{\mathbf{p}}_2, \lambda_2 | J_\mu(0) | -\underset{\sim}{\mathbf{p}}_1, \lambda_1 \rangle^* \quad ; \ \text{time-reversal}$$

Now make use of the covariant form to obtain

$$(\mathcal{J}_\mu)_{fi} = -S_{\lambda_1}S_{\lambda_2} \left\{ i\bar{\mathcal{U}}(-\mathbf{p}_2, \lambda_2) \right.$$
$$\left. \times \left[F_1(q^2)\gamma_\mu - F_2(q^2)\sigma_{\mu\nu}q_\nu + iF_3(q^2)\gamma_5\sigma_{\mu\nu}q_\nu \right] \mathcal{U}(-\mathbf{p}_1, \lambda_1) \right\}^\star$$
$$; \ q = [-\mathbf{p}_2 + \mathbf{p}_1, i(p_{20} - p_{10})]$$

The time-reversed Dirac spinors satisfy the relations in Eqs. (8.75)

$$\mathcal{U}(-\mathbf{p}, \lambda)^* = i\mathcal{S}_\lambda \gamma_1 \gamma_3 \mathcal{U}(\mathbf{p}, \lambda)$$

Hence

$$(\mathcal{J}_\mu)_{fi} = i\bar{\mathcal{U}}(\mathbf{p}_2, \lambda_2)$$
$$\times S_t^{-1} \left[F_1(q^2)\gamma_\mu^* - F_2(q^2)\sigma_{\mu\nu}^* q_\nu^* - iF_3(q^2)\gamma_5 \sigma_{\mu\nu}^* q_\nu^* \right] S_t \mathcal{U}(\mathbf{p}_1, \lambda_1)$$
$$; \ q = [-\mathbf{p}_2 + \mathbf{p}_1, \ i(p_{20} - p_{10})]$$

where we have used the fact from part (b) that the form factors F_i are real, as are (γ_4, γ_5). Furthermore, S_t is the Dirac time-reversal matrix in Eqs. (8.77)–(8.79) with the property

$$S_t = \gamma_1 \gamma_3$$
$$S_t^{-1} \gamma_\mu^* S_t = \gamma_\mu$$

Recall that $\sigma_{\mu\nu} = [\gamma_\mu, \gamma_\nu]/2i$. Thus one also has

$$S_t^{-1} \sigma_{\mu\nu}^* q_\nu^* S_t \rightarrow \sigma_{\mu\nu} q_\nu \qquad ; \ q = p_2 - p_1$$

where, now, $q = p_2 - p_1$. Therefore

$$(\mathcal{J}_\mu)_{fi} = i\bar{\mathcal{U}}(\mathbf{p}_2, \lambda_2) \left[F_1(q^2)\gamma_\mu - F_2(q^2)\sigma_{\mu\nu} q_\nu - iF_3(q^2)\gamma_5 \sigma_{\mu\nu} q_\nu \right] \mathcal{U}(\mathbf{p}_1, \lambda_1)$$
$$; \ q = p_2 - p_1$$

A comparison with the original expression

$$(\mathcal{J}_\mu)_{fi} = i\bar{\mathcal{U}}(\mathbf{p}_2 \lambda_2) \left[F_1(q^2)\gamma_\mu - F_2(q^2)\sigma_{\mu\nu} q_\nu + iF_3(q^2)\gamma_5 \sigma_{\mu\nu} q_\nu \right] \mathcal{U}(\mathbf{p}_1 \lambda_1)$$
$$; \ q = p_2 - p_1$$

again leads to the conclusion that F_3 must vanish

$$F_3(q^2) = 0 \qquad ; \ \text{time-reversal}$$

(e) To make a non-relativistic reduction of the term proportional to $F_3(q^2)$ in the current matrix element, we first note that the energy transfer $q_0 = (E_2 - E_1)/\hbar c$ is of order $(1/M)$, and we neglect it[10]

$$q_0 = O\left(\frac{1}{M}\right)$$

---

[10]Note that the cancelling square-roots in S_{fi} should really have $E/\hbar c$.

The non-relativistic limit (NRL) of the Dirac spinors is

$$\mathcal{U}(\mathbf{p}, \lambda) = \begin{pmatrix} \chi_\lambda \\ \boldsymbol{\sigma} \cdot \mathbf{p}\, \chi_\lambda/2M \end{pmatrix} \qquad ; \text{NRL}$$

It is easy to convince oneself that $\mathcal{J}_{fi}^{(3)} = O(1/M)$. The leading contribution therefore comes from the fourth component of $(\mathcal{J}_\mu)_{fi}$, where the term of interest is

$$\rho_{fi}^{(3)} = \bar{\mathcal{U}}(\mathbf{p}_2\lambda_2)\left[iF_3(\mathbf{q}^2)\gamma_5\sigma_{4j}q_j\right]\mathcal{U}(\mathbf{p}_1\lambda_1)$$

The required product of Dirac matrices is worked out as

$$i\gamma_4\gamma_5\frac{1}{i}\gamma_4\gamma_j q_j = -\begin{pmatrix} 0 & -1 \\ -1 & 0 \end{pmatrix}\begin{pmatrix} 0 & -i\boldsymbol{\sigma}\cdot\mathbf{q} \\ i\boldsymbol{\sigma}\cdot\mathbf{q} & 0 \end{pmatrix} = \begin{pmatrix} i\boldsymbol{\sigma}\cdot\mathbf{q} & 0 \\ 0 & -i\boldsymbol{\sigma}\cdot\mathbf{q} \end{pmatrix}$$

Hence, in the NRL,

$$\rho_{fi}^{(3)} = F_3(\mathbf{q}^2)\chi_{\lambda_2}^\dagger\left[i\boldsymbol{\sigma}\cdot\mathbf{q}\right]\chi_{\lambda_1} + O\left(\frac{1}{M}\right) \qquad ; \mathbf{q} = \mathbf{p}_2 - \mathbf{p}_1$$

$$\mathbf{j}_{fi}^{(3)} = O\left(\frac{1}{M}\right) \qquad\qquad\qquad ; \text{NRL}$$

where $\mathbf{q} = \mathbf{p}_2 - \mathbf{p}_1$.

Suppose the nucleon is scattered from a slowly-varying, time-independent external electrostatic potential $\Phi^{\text{ext}}(\mathbf{x})$. The contribution to the scattering amplitude from the $F_3(\mathbf{q}^2)$ term is proportional to[11]

$$f_{fi}^{(3)} = e_0\, \rho_{fi}^{(3)} \int d^3x\, e^{i(\mathbf{p}_1 - \mathbf{p}_2)\cdot\mathbf{x}}\, \Phi^{\text{ext}}(\mathbf{x})$$

$$= e_0\, \rho_{fi}^{(3)}\, \tilde{\Phi}^{\text{ext}}(\mathbf{q})$$

where $\tilde{\Phi}^{\text{ext}}(\mathbf{q})$ is the Fourier transform of the external potential. Now perform a partial integration, to obtain

$$i\mathbf{q}\tilde{\Phi}^{\text{ext}}(\mathbf{q}) = \int d^3x\, \left(-\boldsymbol{\nabla}e^{-i\mathbf{q}\cdot\mathbf{x}}\right)\Phi^{\text{ext}}(\mathbf{x})$$

$$= \int d^3x\, e^{-i\mathbf{q}\cdot\mathbf{x}}\, \boldsymbol{\nabla}\Phi^{\text{ext}}(\mathbf{x})$$

$$= -\tilde{\mathbf{E}}^{\text{ext}}(\mathbf{q})$$

---

[11] See the solutions to Probs. 7.8 and I.1 in [Amore and Walecka (2013)].

Here $\tilde{\mathbf{E}}^{\text{ext}}(\mathbf{q})$ is the Fourier transform of the external electric field. Hence $f_{fi}^{(3)}$ takes the form

$$f_{fi}^{(3)} = -e_0 F_3(\mathbf{q}^2) \chi_{\lambda_2}^\dagger \left[\boldsymbol{\sigma} \cdot \tilde{\mathbf{E}}^{\text{ext}}(\mathbf{q}) \right] \chi_{\lambda_1}$$

This is an electric-dipole interaction with the nucleon's spin, which, as we have seen, violates both parity and time-reversal invariance.

Chapter 10

Feynman Rules for QCD

Problem 10.1 The Gell-Mann matrices $\underline{\lambda}^a$ for $a = 1, 2, \cdots, 8$ are, in order,

$$\begin{pmatrix} & 1 & \\ 1 & & \\ & & \end{pmatrix} \begin{pmatrix} & -i & \\ i & & \\ & & \end{pmatrix} \begin{pmatrix} 1 & & \\ & -1 & \\ & & \end{pmatrix} \begin{pmatrix} & & 1 \\ & & \\ 1 & & \end{pmatrix} \begin{pmatrix} & & -i \\ & & \\ i & & \end{pmatrix}$$

$$\begin{pmatrix} & & \\ & & 1 \\ & 1 & \end{pmatrix} \begin{pmatrix} & & \\ & & -i \\ & i & \end{pmatrix} \begin{pmatrix} 1/\sqrt{3} & & \\ & 1/\sqrt{3} & \\ & & -2/\sqrt{3} \end{pmatrix}$$

The corresponding non-zero structure constants are (recall they are anti-symmetric in the indices)

$$f^{123} = 1$$

$$f^{147} = -f^{156} = f^{246} = f^{257} = f^{345} = -f^{367} = \frac{1}{2}$$

$$f^{458} = f^{678} = \frac{\sqrt{3}}{2}$$

Pick any subset of these relations and verify them by explicit multiplication of the matrices involved.

Solution to Problem 10.1

Consider, for example,

$$\left[\frac{1}{2}\lambda^1,\frac{1}{2}\lambda^2\right] = \frac{1}{4}\begin{pmatrix}0&1&0\\1&0&0\\0&0&0\end{pmatrix}\begin{pmatrix}0&-i&0\\i&0&0\\0&0&0\end{pmatrix} - \frac{1}{4}\begin{pmatrix}0&-i&0\\i&0&0\\0&0&0\end{pmatrix}\begin{pmatrix}0&1&0\\1&0&0\\0&0&0\end{pmatrix}$$

$$= \frac{i}{2}\begin{pmatrix}1&0&0\\0&-1&0\\0&0&0\end{pmatrix} = \frac{i}{2}\lambda^3$$

Therefore

$$f^{123} = 1$$

Take another example

$$\left[\frac{1}{2}\lambda^4,\frac{1}{2}\lambda^5\right] = \frac{1}{4}\begin{pmatrix}0&0&1\\0&0&0\\1&0&0\end{pmatrix}\begin{pmatrix}0&0&-i\\0&0&0\\i&0&0\end{pmatrix} - \frac{1}{4}\begin{pmatrix}0&0&-i\\0&0&0\\i&0&0\end{pmatrix}\begin{pmatrix}0&0&1\\0&0&0\\1&0&0\end{pmatrix}$$

$$= \frac{i}{2}\begin{pmatrix}1&0&0\\0&0&0\\0&0&-1\end{pmatrix}$$

$$= \frac{i}{4}\begin{pmatrix}1&0&0\\0&-1&0\\0&0&0\end{pmatrix} + \frac{i\sqrt{3}}{4}\begin{pmatrix}1/\sqrt{3}&0&0\\0&1/\sqrt{3}&0\\0&0&-2/\sqrt{3}\end{pmatrix}$$

$$= \frac{i}{4}\lambda^3 + \frac{i\sqrt{3}}{4}\lambda^8$$

Therefore

$$f^{453} = \frac{1}{2} \qquad ; f^{458} = \frac{\sqrt{3}}{2}$$

The complete set of structure constants is evaluated in the solution to Prob. B.5 in [Amore and Walecka (2013)]. The calculated structure constants are antisymmetric in the three indices and invariant under cyclic permutations of the indices.

Problem 10.2 (a) Use the quark field ψ from Eqs. (10.1)–(10.2), and the matrices $\underline{\lambda}^a$ from Prob. 10.1, to write out in detail the Yukawa coupling of quarks to gluons in the second of Eqs. (10.23). Leave the Dirac matrix products intact;

(b) Repeat for the equations of motion in Eq. (10.27).

Solution to Problem 10.2

(a) With the use of the Gell-Mann matrices given in Prob. 10.1, we write[1]

$$\underline{\lambda}^a A^a_\mu = \begin{pmatrix} A^3_\mu + A^8_\mu/\sqrt{3} & A^1_\mu - iA^2_\mu & A^4_\mu - iA^5_\mu \\ A^1_\mu + iA^2_\mu & -A^3_\mu + A^8_\mu/\sqrt{3} & A^6_\mu - iA^7_\mu \\ A^4_\mu + iA^5_\mu & A^6_\mu + iA^7_\mu & -2A^8_\mu/\sqrt{3} \end{pmatrix}$$

Now use Eq. (10.2) to write the quark field as a three-component vector in color space

$$\underline{\psi} = \begin{pmatrix} \psi_R \\ \psi_G \\ \psi_B \end{pmatrix}$$

The Yukawa coupling of the quarks to gluons is contained in $\mathcal{L}_1$ and then reads in detail

$$\frac{ig}{2}\,\underline{\bar{\psi}}\gamma_\mu\underline{\lambda}^a\underline{\psi}A^a_\mu = \frac{ig}{2}\left\{ \left(A^3_\mu + \frac{A^8_\mu}{\sqrt{3}}\right)\bar{\psi}_R\gamma_\mu\psi_R + \left(-A^3_\mu + \frac{A^8_\mu}{\sqrt{3}}\right)\bar{\psi}_G\gamma_\mu\psi_G \right.$$

$$-\frac{2A^8_\mu}{\sqrt{3}}\bar{\psi}_B\gamma_\mu\psi_B + \left(A^1_\mu - iA^2_\mu\right)\bar{\psi}_R\gamma_\mu\psi_G + \left(A^1_\mu + iA^2_\mu\right)\bar{\psi}_G\gamma_\mu\psi_R$$

$$+ \left(A^4_\mu - iA^5_\mu\right)\bar{\psi}_R\gamma_\mu\psi_B + \left(A^4_\mu + iA^5_\mu\right)\bar{\psi}_B\gamma_\mu\psi_R$$

$$\left. + \left(A^6_\mu - iA^7_\mu\right)\bar{\psi}_G\gamma_\mu\psi_B + \left(A^6_\mu + iA^7_\mu\right)\bar{\psi}_B\gamma_\mu\psi_G \right\}$$

(b) The equation of motion (10.27) for massless quarks in QCD reads

$$\gamma_\mu\left[\frac{\partial}{\partial x_\mu} - \frac{ig}{2}\underline{\lambda}^a A^a_\mu(x)\right]\underline{\psi} = 0$$

With the explicit form of $\underline{\lambda}^a A^a_\mu(x)$ reported in part (a), we have

$$\gamma_\mu\left\{ \frac{\partial}{\partial x_\mu}\begin{pmatrix}\psi_R \\ \psi_G \\ \psi_B\end{pmatrix} - \frac{ig}{2} \right.$$

$$\times \left. \begin{bmatrix} \left(A^3_\mu + \frac{A^8_\mu}{\sqrt{3}}\right)\psi_R + (A^1_\mu - iA^2_\mu)\psi_G + (A^4_\mu - iA^5_\mu)\psi_B \\ (A^1_\mu + iA^2_\mu)\psi_R + \left(\frac{A^8_\mu}{\sqrt{3}} - A^3_\mu\right)\psi_G + (A^6_\mu - iA^7_\mu)\psi_B \\ (A^4_\mu + iA^5_\mu)\psi_R + (A^6_\mu + iA^7_\mu)\psi_G - \frac{2A^8_\mu}{\sqrt{3}}\psi_B \end{bmatrix} \right\} = 0$$

---

[1] Recall that here the repeated color index a is summed from 1 to 8.

Recall from Eqs. (10.1)–(10.2) that each color field contains various flavors of quarks [here (u,d,s,c)], and each of these, in turn, is a four-component Dirac field [see Eqs. (10.3)–(10.4)]

$$\underline{\psi} = \begin{pmatrix} \psi_R \\ \psi_G \\ \psi_B \end{pmatrix} \quad ; \; \psi_R = \begin{pmatrix} u_R \\ d_R \\ s_R \\ c_R \end{pmatrix} \quad ; \; u_R = \begin{pmatrix} u_{R1} \\ u_{R2} \\ u_{R3} \\ u_{R4} \end{pmatrix} \quad ; \; etc.$$

Thus this set of equations of motion, when written out in detail, actually represents 12 coupled, linear equations if the Dirac matrix products are left intact.

To see the structure of these equations, we write them out explicitly in the *nuclear domain* of (u, d) quarks

$$\gamma_\mu \left\{ \frac{\partial u_R}{\partial x_\mu} - \frac{ig}{2} \left[\left(A_\mu^3 + \frac{A_\mu^8}{\sqrt{3}} \right) u_R + (A_\mu^1 - iA_\mu^2)u_G + (A_\mu^4 - iA_\mu^5)u_B \right] \right\} = 0$$

$$\gamma_\mu \left\{ \frac{\partial d_R}{\partial x_\mu} - \frac{ig}{2} \left[\left(A_\mu^3 + \frac{A_\mu^8}{\sqrt{3}} \right) d_R + (A_\mu^1 - iA_\mu^2)d_G + (A_\mu^4 - iA_\mu^5)d_B \right] \right\} = 0$$

$$\gamma_\mu \left\{ \frac{\partial u_G}{\partial x_\mu} - \frac{ig}{2} \left[(A_\mu^1 + iA_\mu^2) u_R + \left(\frac{A_\mu^8}{\sqrt{3}} - A_\mu^3 \right) u_G + (A_\mu^6 - iA_\mu^7)u_B \right] \right\} = 0$$

$$\gamma_\mu \left\{ \frac{\partial d_G}{\partial x_\mu} - \frac{ig}{2} \left[(A_\mu^1 + iA_\mu^2) d_R + \left(\frac{A_\mu^8}{\sqrt{3}} - A_\mu^3 \right) d_G + (A_\mu^6 - iA_\mu^7)d_B \right] \right\} = 0$$

$$\gamma_\mu \left\{ \frac{\partial u_B}{\partial x_\mu} - \frac{ig}{2} \left[(A_\mu^4 + iA_\mu^5) u_R + (A_\mu^6 + iA_\mu^7) u_G - \frac{2A_\mu^8}{\sqrt{3}}u_B \right] \right\} = 0$$

$$\gamma_\mu \left\{ \frac{\partial d_B}{\partial x_\mu} - \frac{ig}{2} \left[(A_\mu^4 + iA_\mu^5) d_R + (A_\mu^6 + iA_\mu^7) d_G - \frac{2A_\mu^8}{\sqrt{3}}d_B \right] \right\} = 0$$

Several comments:

- The theory is designed to be invariant under a local color SU(3) transformation, whose infinitesimal form is given in Eqs. (10.15);
- Quarks of a given flavor are coupled only to quarks of the same flavor; the strong interaction does not change the flavor of the quarks;
- The (u, d) quarks indeed have a very small mass, so the absence of a mass term is appropriate here;
- The (u, d) quarks enter in an entirely equivalent fashion;

- Define the following transformation of the two-component fields with (u, d) quarks

$$\psi_R \to e^{i\boldsymbol{\omega}\cdot\boldsymbol{\tau}/2}\,\psi_R \qquad ; \; \psi_G \to e^{i\boldsymbol{\omega}\cdot\boldsymbol{\tau}/2}\,\psi_G \qquad ; \; \psi_B \to e^{i\boldsymbol{\omega}\cdot\boldsymbol{\tau}/2}\,\psi_B$$

which is equivalent to

$$\underline{\psi} \to e^{i\boldsymbol{\omega}\cdot\boldsymbol{\tau}/2}\,\underline{\psi}$$

where the transformation acts in the flavor space of the quark field. The equations of motion are then invariant under this flavor SU(2) transformation. This is the *isospin-invariance* of the strong interaction.[2]

- If the previous equations of motion are extended to include massless strange quarks, then those equations are invariant under a flavor SU(3) transformation; this leads to *The Eightfold Way* of [Gell-Mann and Ne'eman (1964)];

- Although not evident from these field equations, strong vacuum polarization completely shields the color charge, and the states of the observed hadrons are color singlets.

Problem 10.3 Start from the equations of motion for QCD in Eqs. (10.27), (10.29), (10.31), and show that each of the currents in Eqs. (10.32)–(10.34) is *conserved*.

Solution to Problem 10.3

The equations of motion for the massless quark and gluon fields in Eqs. (10.27)–(10.31) are

$$\gamma_\mu \left[\frac{\partial}{\partial x_\mu} - \frac{ig}{2}\lambda^a A_\mu^a(x) \right] \underline{\psi} = 0$$

$$\underline{\bar\psi}\gamma_\mu \left[\frac{\overleftarrow{\partial}}{\partial x_\mu} + \frac{ig}{2}\lambda^a A_\mu^a(x) \right] = 0$$

$$\frac{\partial}{\partial x_\mu}\mathcal{F}_{\nu\mu}^a = \frac{ig}{2}\underline{\bar\psi}\gamma_\nu\lambda^a\underline{\psi} + g f^{abc}\mathcal{F}_{\nu\mu}^b A_\mu^c$$

Use these equations of motion to compute the four-divergence of the various currents.

---

[2] The gluon fields are unchanged under this transformation.

- For the baryon current in Eq. (10.32)

$$\frac{\partial}{\partial x_\mu} \left(i\bar{\underline{\psi}}\gamma_\mu\underline{\psi} \right) = i \left(\frac{\partial}{\partial x_\mu}\bar{\underline{\psi}} \right) \gamma_\mu\underline{\psi} + i\bar{\underline{\psi}}\gamma_\mu \left(\frac{\partial}{\partial x_\mu}\underline{\psi} \right)$$

$$= \frac{g}{2}\bar{\underline{\psi}}\gamma_\mu\lambda^a A_\mu^a(x)\underline{\psi} - \frac{g}{2}\bar{\underline{\psi}}\gamma_\mu\lambda^a A_\mu^a(x)\underline{\psi}$$

This clearly vanishes

$$\frac{\partial}{\partial x_\mu} \left(i\bar{\underline{\psi}}\gamma_\mu\underline{\psi} \right) = 0$$

- For the color current in Eq. (10.33)

$$\frac{\partial}{\partial x_\mu} \left(\frac{i}{2}\bar{\underline{\psi}}\gamma_\mu\lambda^a\underline{\psi} + f^{abc}\mathcal{F}_{\mu\nu}^b A_\nu^c \right) = \frac{\partial}{\partial x_\mu} \left(\frac{\partial}{\partial x_\nu}\mathcal{F}_{\mu\nu}^a \right)$$

$$= \frac{\partial^2}{\partial x_\mu\partial x_\nu}\mathcal{F}_{\mu\nu}^a$$

Since $\mathcal{F}_{\mu\nu}^a = -\mathcal{F}_{\nu\mu}^a$ is antisymmetric, this vanishes identically

$$\frac{\partial}{\partial x_\mu} \left(\frac{i}{2}\bar{\underline{\psi}}\gamma_\mu\lambda^a\underline{\psi} + f^{abc}\mathcal{F}_{\mu\nu}^b A_\nu^c \right) = 0$$

- For the flavor current in Eq. (10.34)

$$\frac{\partial}{\partial x_\mu} \left(i\bar{\underline{\psi}}\gamma_\mu\Sigma\,\underline{\psi} \right) = i \left(\frac{\partial}{\partial x_\mu}\bar{\underline{\psi}} \right) \gamma_\mu\Sigma\underline{\psi} + i\bar{\underline{\psi}}\Sigma\gamma_\mu \left(\frac{\partial}{\partial x_\mu}\underline{\psi} \right)$$

$$= \frac{g}{2}\bar{\underline{\psi}}\gamma_\mu\lambda^a A_\mu^a(x)\Sigma\underline{\psi} - \frac{g}{2}\bar{\underline{\psi}}\Sigma\gamma_\mu\lambda^a A_\mu^a(x)\underline{\psi}$$

The first line follows since the matrix Σ acts in the flavor space; it has nothing to do with the Dirac indices.

Since Σ acts in the flavor space (it is effectively the unit matrix with respect to color), and $\underline{\lambda}^a$ acts in the color space (it is effectively the unit matrix with respect to flavor), and neither have anything to do with the Dirac indices, this expression is the same as[3]

$$\frac{\partial}{\partial x_\mu} \left(i\bar{\underline{\psi}}\gamma_\mu\Sigma\underline{\psi} \right) = \frac{g}{2}\bar{\underline{\psi}}\lambda^a A_\mu^a(x)\gamma_\mu\Sigma\,\underline{\psi} - \frac{g}{2}\bar{\underline{\psi}}\lambda^a A_\mu^a(x)\gamma_\mu\Sigma\,\underline{\psi}$$

---

[3]Compare the following problem. Note that the addition of a quark mass term does not affect the conservation of the baryon and color currents; however, the flavor current will only remain conserved if $[\Sigma, M] = 0$.

which clearly also vanishes

$$\frac{\partial}{\partial x_\mu}\left(i\underline{\bar\psi}\gamma_\mu\Sigma\,\underline{\psi}\right)=0$$

Problem 10.4 A more explicit notation is to write the quark wave functions as the direct product of a three-component color wave function, a four-component flavor wave function, and a four-component Dirac spinor

$$[\eta^{\text{color}}\otimes\zeta^{\text{flavor}}\otimes u^{\text{Dirac}}]_{ilr}=\eta_i\zeta_l u_r \qquad\text{; quark wave functions}$$

Matrices are then the direct product of matrices in each space

$$[a\otimes b\otimes c]_{ilr,jms}=a_{ij}b_{lm}c_{rs}\qquad\text{; direct product}$$

(a) Write the Dirac current in Eq. (10.33) in terms of the matrix $[\lambda\otimes 1\otimes\gamma_\mu]$;

(b) Write Eq. (10.16) in terms of the matrix $[1\otimes M\otimes 1]$;

(c) Write Eq. (10.34) in terms of the matrix $[1\otimes\Sigma\otimes\gamma_\mu]$.

Unit matrices and direct-product symbols are conventionally suppressed.[4]

Solution to Problem 10.4

Express the wave functions in the quark field expansions as $[\eta^{\text{color}}\otimes\zeta^{\text{flavor}}\otimes u^{\text{Dirac}}]_{ilr}$. The field will then have a corresponding triplet of indices ψ_{ilr}, which get summed over in matrix products.

(a) The representation of the Dirac color current in Eq. (10.33) is then

$$\frac{i}{2}\,\underline{\bar\psi}\gamma_\mu\lambda^a\underline{\psi}\to\frac{i}{2}\bar\psi\,[\lambda^a\otimes 1\otimes\gamma_\mu]\,\psi$$

(b) The quark mass term in Eq. (10.16) takes the form

$$-\underline{\bar\psi}M\underline{\psi}\to-\bar\psi\,[1\otimes M\otimes 1]\,\psi$$

(c) The flavor current in Eq. (10.34) becomes

$$i\underline{\bar\psi}\gamma_\mu\Sigma\,\underline{\psi}\to i\bar\psi\,[1\otimes\Sigma\otimes\gamma_\mu]\,\psi$$

Problem 10.5 (a) Show the canonical momenta conjugate to the gluon

---

[4]The order of display of the matrices $[a\otimes b\otimes c]$ in the text is a matter of convenience.

field A_j^a in QCD are

$$\Pi_j^a \equiv \Pi_{A_j^a} = i\mathcal{F}_{4j}^a \qquad ; \text{ canonical momenta}$$

$$; j = 1, 2, 3 \qquad ; a = 1, \cdots, 8$$

(b) Use Eq. (10.31) to derive Gauss's law for the color fields

$$\nabla \cdot \Pi^a = -\frac{g}{2}\psi^\dagger \underline{\lambda}^a \underline{\psi} + gf^{abc}\,\Pi^b \cdot \mathbf{A}^c$$

$$\equiv -g(\rho_{\text{quark}}^a + \rho_{\text{gluon}}^a) \qquad ; \text{ Gauss's law}$$

Solution to Problem 10.5

(a) Equations (10.28) and (10.30) for derivatives of the lagrangian density with respect to derivatives of the fields read

$$\frac{\partial \mathcal{L}}{\partial(\partial\underline{\psi}/\partial x_\mu)} = -\bar{\underline{\psi}}\gamma_\mu$$

$$\frac{\partial \mathcal{L}}{\partial(\partial A_\nu^a/\partial x_\mu)} = \mathcal{F}_{\nu\mu}^a$$

The momentum density conjugate to the coordinate q is given in lagrangian field theory by

$$\Pi_q = \frac{\partial \mathcal{L}}{\partial(\partial q/\partial t)}$$

If we use $x_4 = it$, then it follows from the above that[5]

$$\underline{\Pi}_\psi = \frac{\partial \mathcal{L}}{\partial(\partial\underline{\psi}/\partial t)} = i\underline{\psi}^\dagger$$

$$\Pi_{A_\nu^a} = \frac{\partial \mathcal{L}}{\partial(\partial A_\nu^a/\partial t)} = i\mathcal{F}_{4\nu}^a$$

Note that the temporal component of the last relation, providing the momentum conjugate to A_4^a, vanishes identically since $\mathcal{F}_{44}^a = 0$[6]

$$\Pi_{A_4^a} \equiv 0$$

Thus A_4^a is not an independent dynamical variable.

---

[5]Recall that in chapter 10 we simplify to units where $\hbar = c = 1$.
[6]Note that $\Pi_{\bar{\psi}} = 0$ also vanishes.

The answer to part (a) is obtained by writing the non-vanishing spatial components of the gluon canonical momentum density as

$$\Pi_j^a \equiv \frac{\partial \mathcal{L}}{\partial(\partial A_j^a/\partial t)} = i\mathcal{F}_{4j}^a$$

(b) The equation of motion for $\mathcal{F}_{\nu\mu}^a$ is [see Eq. (10.31)]

$$\frac{\partial}{\partial x_\mu}\mathcal{F}_{\nu\mu}^a = \frac{ig}{2}\bar{\psi}\gamma_\nu\lambda^a\psi + gf^{abc}\mathcal{F}_{\nu\mu}^b A_\mu^c$$

Therefore, since again $\mathcal{F}_{44}^a = 0$,

$$\frac{\partial}{\partial x_j}\mathcal{F}_{4j}^a = \frac{ig}{2}\bar{\psi}\gamma_4\lambda^a\psi + gf^{abc}\mathcal{F}_{4j}^b A_j^c$$

In vector notation this reads[7]

$$\boldsymbol{\nabla}\cdot\boldsymbol{\Pi}^a = -\frac{g}{2}\psi^\dagger\lambda^a\psi + gf^{abc}\,\boldsymbol{\Pi}^b\cdot\mathbf{A}^c$$

$$\equiv -g\left(\rho_{\text{quark}}^a + \rho_{\text{gluon}}^a\right) \qquad ;\text{ Gauss's law}$$

This is Gauss's law for the gluon field, where the second line defines the quark and gluon color charge densities. Note that

- Since there is no time derivative in this relation, *it is not an operator equation of motion, rather, it provides a constraint equation on* $\boldsymbol{\Pi}^a$ *which must be satisfied at all times;*
- In the source term, the gluon color charge density involves a non-linear combination of gluon fields, again depending on $\boldsymbol{\Pi}^a$.

Problem 10.6 (a) Show, with the aid of partial integration, that the hamiltonian density in QCD can be written

$$\mathcal{H}_{\text{QCD}} \doteq \frac{1}{2}\boldsymbol{\Pi}^a\cdot\boldsymbol{\Pi}^a + \frac{1}{4}\mathcal{F}_{ij}^a\mathcal{F}_{ij}^a + \psi^\dagger\boldsymbol{\alpha}\cdot\left[\frac{1}{i}\boldsymbol{\nabla} - \frac{g}{2}\lambda^a\mathbf{A}^a\right]\psi$$

(b) Prove that any vector field $\boldsymbol{\Pi}$ can be separated into $\boldsymbol{\Pi}_T + \boldsymbol{\Pi}_L$ where[8]

$$\boldsymbol{\nabla}\cdot\boldsymbol{\Pi}_T = 0 \qquad ;\boldsymbol{\nabla}\times\boldsymbol{\Pi}_L = 0 \qquad ;\int_{\text{Box}} d^3x\,\boldsymbol{\Pi}_L\cdot\boldsymbol{\Pi}_T = 0$$

---

[7]We remind the reader that bold face here denotes a three-vector in real space.
[8]See [Fetter and Walecka (2003a)].

(c) Hence, show the hamiltonian density can be written

$$\mathcal{H}_{QCD} \doteq \psi^\dagger \boldsymbol{\alpha} \cdot \left[\frac{1}{i} \boldsymbol{\nabla} - \frac{g}{2} \lambda^a \mathbf{A}^a \right] \psi + \frac{1}{2} \boldsymbol{\Pi}_T^a \cdot \boldsymbol{\Pi}_T^a + \frac{1}{4} \mathcal{F}_{ij}^a \mathcal{F}_{ij}^a$$
$$+ \frac{g^2}{8\pi} \int d^3x \int d^3x' \, \rho^a(\mathbf{x}) \frac{1}{|\mathbf{x} - \mathbf{x}'|} \rho^a(\mathbf{x}')$$

; hamiltonian density

Here $\boldsymbol{\nabla} \cdot \boldsymbol{\Pi}^a = \boldsymbol{\nabla} \cdot \boldsymbol{\Pi}_L^a$ satisfies the constraint equation in Prob. 10.5(b), whose solution in terms of the color charge $\rho^a = \rho_{\text{quark}}^a + \rho_{\text{gluon}}^a$ at any instant in time is

$$\boldsymbol{\Pi}_L^a(\mathbf{x}) = \frac{g}{4\pi} \boldsymbol{\nabla} \int d^3x' \frac{1}{|\mathbf{x} - \mathbf{x}'|} \rho^a(\mathbf{x}')$$

Since ρ_{gluon}^a depends on $\boldsymbol{\Pi}_L$, this is an integral equation (or power series) for $\boldsymbol{\Pi}_L^a$.[9]

Solution to Problem 10.6

(a) With quark coordinates $(\psi, \boldsymbol{\Pi}_\psi)$ and gluon coordinates $(\mathbf{A}^a, \boldsymbol{\Pi}^a)$, where the canonical momentum densities are developed in Prob. 10.5, the hamiltonian density of QCD is given by[10]

$$\mathcal{H} = \boldsymbol{\Pi}^a \cdot \dot{\mathbf{A}}^a + \underline{\boldsymbol{\Pi}}_\psi \dot{\underline{\psi}} - \mathcal{L}$$

Introduce the definition

$$A_\mu^a \equiv (\mathbf{A}^a, i\Phi^a)$$

and write out

$$\Pi_j^a = i\mathcal{F}_{4j}^a = i \left[\frac{\partial}{\partial x_4} A_j^a - \frac{\partial}{\partial x_j} A_4^a + g f^{abc} A_4^b A_j^c \right]$$

In vector notation, this provides the relation between the time derivative $\dot{\mathbf{A}}^a$ and the canonical momentum density $\Pi_{\mathbf{A}^a} \equiv \boldsymbol{\Pi}^a$

$$\boldsymbol{\Pi}^a = \dot{\mathbf{A}}^a + \boldsymbol{\nabla}\Phi^a - g f^{abc} \Phi^b \mathbf{A}^c$$

---

[9]Note that, ultimately, one has an expression $\mathcal{H}_{QCD} = \mathcal{H}_{QCD}(\mathbf{A}^a, \boldsymbol{\Pi}_T^a)$.
[10]Recall that $\Pi_{A_4^a} = \underline{\boldsymbol{\Pi}}_{\bar\psi} = 0$.

Now insert the lagrangian density in Eq. (10.14), and evaluate the hamiltonian density

$$\mathcal{H} = \mathbf{\Pi}^a \cdot \left[\mathbf{\Pi}^a - \boldsymbol{\nabla}\Phi^a + g f^{abc}\Phi^b \mathbf{A}^c \right] + i\underline{\psi}^\dagger \frac{\partial}{\partial t} \underline{\psi}$$

$$- \left[-\underline{\psi}^\dagger \frac{\partial}{\partial it} \underline{\psi} - \underline{\bar{\psi}}\boldsymbol{\gamma} \cdot \boldsymbol{\nabla}\underline{\psi} + \frac{i}{2}g\,\underline{\bar{\psi}}\gamma_\mu \underline{\lambda}^a \underline{\psi} A^a_\mu - \frac{1}{4}\mathcal{F}^a_{ij}\mathcal{F}^a_{ij} - \frac{1}{2}\mathcal{F}^a_{4i}\mathcal{F}^a_{4i} \right]$$

The terms in $\partial\underline{\psi}/\partial t$ cancel, and recalling that $\Pi^a_j = i\mathcal{F}^a_{4j}$, this becomes

$$\mathcal{H} = \mathbf{\Pi}^a \cdot \left[\mathbf{\Pi}^a - \boldsymbol{\nabla}\Phi^a + g f^{abc}\Phi^b \mathbf{A}^c \right]$$

$$- \left[-\underline{\bar{\psi}}\boldsymbol{\gamma} \cdot \boldsymbol{\nabla}\underline{\psi} + \frac{i}{2}g\,\underline{\bar{\psi}}\gamma_\mu \underline{\lambda}^a \underline{\psi} A^a_\mu - \frac{1}{4}\mathcal{F}^a_{ij}\mathcal{F}^a_{ij} + \frac{1}{2}\mathbf{\Pi}^a \cdot \mathbf{\Pi}^a \right]$$

The hamiltonian is the integral of $\mathcal{H}$ over the quantization volume

$$H = \int_{\text{Box}} d^3x\, \mathcal{H}(\mathbf{x})$$

We can carry out *partial integrations* on the hamiltonian density without changing H; the surface terms are eliminated by the boundary conditions. Thus we can discard any total divergences in $\mathcal{H}$. After a partial integration (indicated by $\doteq$), the hamiltonian density becomes[11]

$$\mathcal{H} \doteq \frac{1}{2}\mathbf{\Pi}^a \cdot \mathbf{\Pi}^a + \frac{1}{4}\mathcal{F}^a_{ij}\mathcal{F}^a_{ij} + \underline{\psi}^\dagger \boldsymbol{\alpha} \cdot \left(\frac{1}{i}\boldsymbol{\nabla} - \frac{g}{2}\underline{\lambda}^a \mathbf{A}^a \right) \underline{\psi}$$

$$+ \frac{g}{2}\underline{\psi}^\dagger \underline{\lambda}^a \underline{\psi}\,\Phi^a + \Phi^a \boldsymbol{\nabla} \cdot \mathbf{\Pi}^a + g f^{abc}\left(\mathbf{\Pi}^a \cdot \Phi^b \mathbf{A}^c \right)$$

Now multiply Gauss's law, as derived in Prob. 10.5(b), by Φ^a. It follows that the *sum of terms in the last line vanishes!* Hence, the hamiltonian density of QCD takes the simple, elegant form

$$\mathcal{H} \doteq \frac{1}{2}\mathbf{\Pi}^a \cdot \mathbf{\Pi}^a + \frac{1}{4}\mathcal{F}^a_{ij}\mathcal{F}^a_{ij} + \underline{\psi}^\dagger \boldsymbol{\alpha} \cdot \left(\frac{1}{i}\boldsymbol{\nabla} - \frac{g}{2}\underline{\lambda}^a \mathbf{A}^a \right) \underline{\psi}$$

The remaining field tensor in $\mathcal{H}$ involves only the spatial components of $\mathbf{A}^a$ and their spatial derivatives

$$\mathcal{F}^a_{ij} = \frac{\partial}{\partial x_i}A^a_j - \frac{\partial}{\partial x_j}A^a_i + g f^{abc}A^b_i A^c_j$$

---

[11]Use $\boldsymbol{\gamma} = i\boldsymbol{\alpha}\beta$.

We note that Φ^a has *disappeared* from $\mathcal{H}$; however, we still have to satisfy the constraint equation

$$\boldsymbol{\nabla} \cdot \boldsymbol{\Pi}^a = -\frac{g}{2}\,\underline{\psi}^\dagger \underline{\lambda}^a \underline{\psi} + g f^{abc}\, \boldsymbol{\Pi}^b \cdot \mathbf{A}^c$$

which implies that *not all components of $\boldsymbol{\Pi}^a$ are independent.*

(b) It is shown in Prob. 7.6 that at a given instant, a vector field can always be separated into longitudinal and transverse parts where

$$\boldsymbol{\Pi}(\mathbf{x}) = \boldsymbol{\Pi}_T(\mathbf{x}) + \boldsymbol{\Pi}_L(\mathbf{x})$$

Here

$$\boldsymbol{\Pi}_T(\mathbf{x}) = \frac{1}{\sqrt{\Omega}} \sum_{\mathbf{k}} \sum_{s=1}^{2} \Pi(\mathbf{k}, s)\, \mathbf{e}_{\mathbf{k}s}\, e^{i\mathbf{k}\cdot\mathbf{x}} \qquad ; \boldsymbol{\nabla} \cdot \boldsymbol{\Pi}_T(\mathbf{x}) = 0$$

$$\boldsymbol{\Pi}_L(\mathbf{x}) = \frac{1}{\sqrt{\Omega}} \sum_{\mathbf{k}} \Pi(\mathbf{k}, 0)\, \mathbf{e}_{\mathbf{k}0}\, e^{i\mathbf{k}\cdot\mathbf{x}} \qquad ; \boldsymbol{\nabla} \times \boldsymbol{\Pi}_L(\mathbf{x}) = 0$$

One can use the orthonormality of the plane waves to compute

$$\int_{\text{Box}} d^3x\, \boldsymbol{\Pi}_L(\mathbf{x}) \cdot \boldsymbol{\Pi}_T(\mathbf{x}) = \sum_{\mathbf{k}} \sum_{s=1}^{2} \Pi(\mathbf{k}, s)\Pi(-\mathbf{k}, 0)\mathbf{e}_{\mathbf{k}s} \cdot \mathbf{e}_{-\mathbf{k}0}$$

Since the transverse and longitudinal unit vectors are orthogonal, this vanishes. Hence

$$\int_{\text{Box}} d^3x\, \boldsymbol{\Pi}_L \cdot \boldsymbol{\Pi}_T = 0$$

(c) Separate the canonical momentum density into its transverse and longitudinal parts

$$\boldsymbol{\Pi}^a = \boldsymbol{\Pi}_T^a + \boldsymbol{\Pi}_L^a$$

Gauss's law in Prob. 10.5 then reads

$$\boldsymbol{\nabla} \cdot \boldsymbol{\Pi}^a = \boldsymbol{\nabla} \cdot \boldsymbol{\Pi}_L^a = -\frac{g}{2}\,\underline{\psi}^\dagger \underline{\lambda}^a \underline{\psi} + g f^{abc}\left(\boldsymbol{\Pi}_L^b + \boldsymbol{\Pi}_T^b\right) \cdot \mathbf{A}^c$$

Just as in electrostatics, this equation can be solved for $\boldsymbol{\Pi}_L^a$ with the use of the Coulomb Green's function

$$\boldsymbol{\nabla} \cdot \boldsymbol{\Pi}_L^a = -g\rho^a$$

$$\boldsymbol{\Pi}_L^a(\mathbf{x}) = \frac{g}{4\pi}\boldsymbol{\nabla} \int \frac{\rho^a(\mathbf{x}')d^3x'}{|\mathbf{x} - \mathbf{x}'|}$$

Thus

$$\boldsymbol{\Pi}_L^a(\mathbf{x}) = \frac{g}{4\pi}\boldsymbol{\nabla}\int \frac{1}{|\mathbf{x}-\mathbf{x}'|}\left[\frac{1}{2}\underline{\psi}^\dagger\underline{\lambda}^a\underline{\psi} - f^{abc}\left(\boldsymbol{\Pi}_L^b + \boldsymbol{\Pi}_T^b\right)\cdot\mathbf{A}^c\right]d^3x'$$

This is an *integral equation* for $\boldsymbol{\Pi}_L$. Since the r.h.s. is explicitly of order g, we can iterate this equation to find $\boldsymbol{\Pi}_L$ as a power series in g

$$\boldsymbol{\Pi}_L^a(\mathbf{x}) = \frac{g}{4\pi}\boldsymbol{\nabla}\int \frac{1}{|\mathbf{x}-\mathbf{x}'|}\left[\frac{1}{2}\underline{\psi}^\dagger\underline{\lambda}^a\underline{\psi} - f^{abc}\,\boldsymbol{\Pi}_T^b\cdot\mathbf{A}^c\right]d^3x' + O(g^2)$$

Now substitute these expressions in the hamiltonian density, and use

$$\int_{\text{Box}} \boldsymbol{\Pi}_L^a \cdot \boldsymbol{\Pi}_T^a = 0$$

$$\int_{\text{Box}} \boldsymbol{\Pi}_L^a \cdot \boldsymbol{\Pi}_L^a = \frac{g^2}{4\pi}\int d^3x \int d^3x'\, \rho^a(\mathbf{x}')\frac{1}{|\mathbf{x}-\mathbf{x}'|}\rho^a(\mathbf{x})$$

where the second relation again follows from a partial integration.

In *summary*, the hamiltonian density of QCD is given by

$$\mathcal{H} \doteq \underline{\psi}^\dagger\boldsymbol{\alpha}\cdot\left(\frac{1}{i}\boldsymbol{\nabla} - \frac{g}{2}\underline{\lambda}^a\mathbf{A}^a\right)\underline{\psi} + \frac{1}{2}\boldsymbol{\Pi}_T^a\cdot\boldsymbol{\Pi}_T^a$$

$$+ \frac{g^2}{8\pi}\int d^3x \int d^3x'\, \rho^a(\mathbf{x}')\frac{1}{|\mathbf{x}-\mathbf{x}'|}\rho^a(\mathbf{x}) + \frac{1}{4}\mathcal{F}_{ij}^a\mathcal{F}_{ij}^a$$

where the color charge density receives a contribution from both quarks and gluons

$$\rho^a(\mathbf{x}) = \frac{1}{2}\underline{\psi}^\dagger\underline{\lambda}^a\underline{\psi} - f^{abc}\left(\boldsymbol{\Pi}_L^b + \boldsymbol{\Pi}_T^b\right)\cdot\mathbf{A}^c$$

and the longitudinal canonical momentum density $\boldsymbol{\Pi}_L$ is the solution to the integral equation

$$\boldsymbol{\Pi}_L^a(\mathbf{x}) = \frac{g}{4\pi}\boldsymbol{\nabla}\int d^3x' \frac{1}{|\mathbf{x}-\mathbf{x}'|}\rho^a(\mathbf{x}')$$

Note that the hamiltonian density now becomes a function of the quark field (and its adjoint), the gluon field, and the transverse gluon canonical momentum density $\mathcal{H}(\underline{\psi}, \mathbf{A}^a, \boldsymbol{\Pi}_T^a)$.

Problem 10.7 The quantization of quantum electrodynamics (QED) starting from lagrangian field theory is discussed in appendix C of volume

II. In the Coulomb gauge, in H-L units ($\varepsilon_0 = 1$), the quantization condition on the E-M field is

$$[A_i(\mathbf{x}, t), \Pi_j(\mathbf{x}', t')]_{t=t'} = i\delta_{ij}^T(\mathbf{x} - \mathbf{x}')$$

$$\Pi_j \doteq \Pi_j^T = \frac{\partial A_j}{\partial t}$$

A representation of the fields satisfying these relations is then provided by Eqs. (C.67), with the quantization condition in Eq. (7.38).[12]

Now use the results of Prob. 10.6, and the analogy to QED, to discuss the quantization of QCD in the Coulomb gauge where $\boldsymbol{\nabla} \cdot \mathbf{A}^a = 0$.

Solution to Problem 10.7

So far, no mention has been made of choice of gauge for the gluon field, and the results in Prob. (10.6) are therefore *gauge invariant*. Since the hamiltonian density $\mathcal{H}(\underline{\psi}, \mathbf{A}^a, \boldsymbol{\Pi}_T^a)$ only depends on the transverse canonical momentum density, quantization of the theory is most conveniently carried out in the Coulomb gauge where the gluon field is also transverse

$$\boldsymbol{\nabla} \cdot \mathbf{A}^a = 0 \qquad\qquad ; \text{ Coulomb gauge}$$

We will not give a general proof that such a gauge can always be found, but rather demonstrate this in the case where a vector potential A_μ^a has been chosen, and the resulting $\boldsymbol{\nabla} \cdot \mathbf{A}^a$ is small. Now go to a gauge-transformed field $A_\mu^{a\prime}$, where the new field is given by the second of Eqs. (10.15)

$$A_\mu^{a\prime} = A_\mu^a - \frac{1}{g}\frac{\partial \theta^a}{\partial x_\mu} + f^{abc}\theta^b A_\mu^c \qquad ; \theta^a \to 0$$

We require that $\boldsymbol{\nabla} \cdot \mathbf{A}^{a\prime}$ vanish

$$\boldsymbol{\nabla} \cdot \mathbf{A}^{a\prime} = \boldsymbol{\nabla} \cdot \mathbf{A}^a - \frac{1}{g}\nabla^2\theta^a + f^{abc}\boldsymbol{\nabla}\theta^b \cdot \mathbf{A}^c + f^{abc}\theta^b\,\boldsymbol{\nabla} \cdot \mathbf{A}^c$$

$$= 0$$

This equation is re-written as

$$\nabla^2\theta^a - \left(gf^{abc}\mathbf{A}^c\right)\cdot\boldsymbol{\nabla}\theta^b - \left(gf^{abc}\,\boldsymbol{\nabla}\cdot\mathbf{A}^c\right)\theta^b = g\boldsymbol{\nabla}\cdot\mathbf{A}^a$$

This presents a set of eight, coupled, linear, inhomogeneous, second-order partial differential equations for θ^a, which can be integrated to find the

---

[12]Compare Eqs. (7.65). See also the hamiltonian density in Eq. (C.68).

appropriate $\theta^a(\mathbf{x})$ at any instant.[13] The new vector potential is then in the Coulomb gauge.

One can now use the analogy with QED to quantize the theory by imposing the following canonical commutation relations on the gluon field in the Coulomb gauge[14]

$$[A_i^a(\mathbf{x},t), \Pi_{Tj}^b(\mathbf{x}',t')]_{t=t'} = i\delta^{ab}\,\delta_{ij}^T(\mathbf{x}-\mathbf{x}')$$
$$\nabla \cdot \mathbf{A}^a(\mathbf{x},t) = 0 \qquad\qquad ; \text{ Coulomb gauge}$$

The quark fields satisfy the canonical anticommutation relations

$$\left\{\psi_\alpha(\mathbf{x},t), \psi_\beta^\dagger(\mathbf{x}',t')\right\}_{t=t'} = \delta_{\alpha\beta}\,\delta^{(3)}(\mathbf{x}-\mathbf{x}')$$

Problem 10.8 Consider a field theory with a Dirac fermion field ψ, a scalar boson field ϕ, and a Yukawa coupling between them

$$\mathcal{L} = \mathcal{L}_0 + \mathcal{L}_1 = -\bar{\psi}\left[\gamma_\mu\frac{\partial}{\partial x_\mu} + (M - g\phi)\right]\psi - \frac{1}{2}\left[\left(\frac{\partial\phi}{\partial x_\mu}\right)^2 + m^2\phi^2\right]$$
$$\mathcal{L}_1 = g\bar{\psi}\psi\phi$$

The generating functional for this theory is[15]

$$\tilde{W}(\bar{\zeta},\zeta,J) = \exp\left\{i\int d^4x\left[g\left(-\frac{1}{i}\frac{\delta}{\delta\zeta(x)}\right)\left(\frac{1}{i}\frac{\delta}{\delta\bar{\zeta}(x)}\right)\left(\frac{1}{i}\frac{\delta}{\delta J(x)}\right)\right]\right\}$$
$$\times \tilde{W}_0(\bar{\zeta},\zeta,J) / [\cdots]_{\zeta=\bar{\zeta}=J=0}$$

where

$$\tilde{W}_0(\bar{\zeta},\zeta,J) = \exp\left\{-i\int\int d^4x d^4y\,\bar{\zeta}(x)S_F(x-y)\zeta(y)\right\}$$
$$\times \exp\left\{\frac{i}{2}\int\int d^4x d^4y\,J(x)\Delta_F(x-y)J(y)\right\}$$

---

[13]For example, if the second and third terms on the l.h.s. were absent, these would reduce to Poisson's equation $\nabla^2\theta^a = g\nabla\cdot\mathbf{A}^a$, which is readily integrated. This proof of the existence of the Coulomb gauge can be extended by considering repeated application of infinitesimals [Amore (2014)].

[14]See Eqs. (C.43) in Vol. II, and the accompanying discussion. For one illustration of the use of the Coulomb gauge in QCD, see [Andraši and Taylor (2012)].

[15]The goal of this problem is to give the reader some proficiency in obtaining Feynman rules from the generating functional (compare appendix C in [Serot and Walecka (1986)]). Here $(\bar{\zeta},\zeta)$ form Grassmann algebras, and the variational derivatives in $\tilde{W}$ are to be interpreted as "left derivatives".

(a) Expand the first exponential in $\tilde{W}$, and evaluate the numerator in $\tilde{W}$ to order g^2;

(b) Use this result to evaluate the denominator in $\tilde{W}$ to order g^2. Show to this order that the denominator in $\tilde{W}$ serves to precisely cancel the disconnected diagrams;

(c) Hence derive an expression to order g^2 for the fermion propagator

$$iS_F'(x-y) = \left[\frac{1}{i}\frac{\delta}{\delta\bar{\zeta}(x)} \tilde{W}(\bar{\zeta},\zeta,J)\frac{1}{i}\frac{\delta}{\delta\zeta(y)} \right]_{\zeta=\bar{\zeta}=J=0}$$

(d) Interpret your result in terms of a set of coordinate-space Feynman rules for this theory.

Solution to Problem 10.8

(a) An expansion of the first exponential in $\tilde{W}(\bar{\zeta},\zeta,J)$ up through order g^2 gives

$$\tilde{W}(\bar{\zeta},\zeta,J) \approx \left\{ 1 + ig\int d^4x \left[-\frac{1}{i}\frac{\delta}{\delta\zeta(x)} \right]\left[\frac{1}{i}\frac{\delta}{\delta\bar{\zeta}(x)} \right]\left[\frac{1}{i}\frac{\delta}{\delta J(x)} \right] \right.$$
$$\left. - \frac{g^2}{2}\left(\int d^4x \left[-\frac{1}{i}\frac{\delta}{\delta\zeta(x)} \right]\left[\frac{1}{i}\frac{\delta}{\delta\bar{\zeta}(x)} \right]\left[\frac{1}{i}\frac{\delta}{\delta J(x)} \right] \right)^2 \right\}$$
$$\times \tilde{W}_0(\bar{\zeta},\zeta,J) / [\cdots]_{\zeta=\bar{\zeta}=J=0}$$

Let us explicitly evaluate the term of $O(g)$ in the numerator

$$\mathcal{D}_1(\bar{\zeta},\zeta,J) \equiv \int d^4z \left[-\frac{1}{i}\frac{\delta}{\delta\zeta(z)} \right]\left[\frac{1}{i}\frac{\delta}{\delta\bar{\zeta}(z)} \right]\left[\frac{1}{i}\frac{\delta}{\delta J(z)} \right] \tilde{W}_0(\bar{\zeta},\zeta,J)$$

This gives

$$\mathcal{D}_1(\bar{\zeta},\zeta,J) = \int d^4z \left[\int d^4x\, J(x)\Delta_F(x-z) \right]\left[-\frac{1}{i}\frac{\delta}{\delta\zeta(z)} \right]\left[\frac{1}{i}\frac{\delta}{\delta\bar{\zeta}(z)} \right]$$
$$\times \tilde{W}_0(\bar{\zeta},\zeta,J)$$
$$= \int d^4z \left[\int d^4x\, J(x)\Delta_F(x-z) \right]\left[-\frac{1}{i}\frac{\delta}{\delta\zeta(z)} \right]$$
$$\times \left[-\int d^4y\, S_F(z-y)\zeta(y) \right] \tilde{W}_0(\bar{\zeta},\zeta,J)$$

Hence[16]

$$\mathcal{D}_1(\bar{\zeta}, \zeta, J) = \int d^4 z \left[\int d^4 x\, J(x)\Delta_F(x-z) \right]$$

$$\times \left[\frac{1}{i}\, S_F(0) + \int d^4 x \int d^4 y\, \bar{\zeta}(x) S_F(x-z)\, S_F(z-y)\zeta(y) \right] \tilde{W}_0(\bar{\zeta}, \zeta, J)$$

We will now use this result to evaluate the term of $O(g^2)$ in the numerator, where we write the square of the integral as a double integral

$$\mathcal{D}_2(\bar{\zeta}, \zeta, J) \equiv \left(\int d^4 z \left[-\frac{1}{i}\frac{\delta}{\delta\zeta(z)} \right] \left[\frac{1}{i}\frac{\delta}{\delta\bar{\zeta}(z)} \right] \left[\frac{1}{i}\frac{\delta}{\delta J(z)} \right] \right)^2 \tilde{W}_0(\bar{\zeta}, \zeta, J)$$

$$= \int d^4 u \left[-\frac{1}{i}\frac{\delta}{\delta\zeta(u)} \right] \left[\frac{1}{i}\frac{\delta}{\delta\bar{\zeta}(u)} \right] \left[\frac{1}{i}\frac{\delta}{\delta J(u)} \right] \mathcal{D}_1(\bar{\zeta}, \zeta, J)$$

It is useful to consider the functional derivative

$$\mathcal{F} \equiv \left[-\frac{1}{i}\frac{\delta}{\delta\zeta(u)} \right] \left[\frac{1}{i}\frac{\delta}{\delta\bar{\zeta}(u)} \right] F(\zeta, \bar{\zeta}, J) G(\zeta, \bar{\zeta}, J)$$

where F and G are functionals of the sources ζ, $\bar{\zeta}$ and J. We can write

$$\mathcal{F} = \frac{\delta}{\delta\zeta(u)} \left[\frac{\delta F}{\delta\bar{\zeta}(u)}\, G + (-1)^{\eta_F} F \frac{\delta G}{\delta\bar{\zeta}(u)} \right]$$

where η_F is the number of Grassmann variables in F (one needs to move the functional derivative $\delta/\delta\zeta(u)$ through these terms, with a sign flip each time that an interchange is performed). Similarly, one can perform the remaining derivative

$$\mathcal{F} = \frac{\delta^2 F}{\delta\zeta(u)\delta\bar{\zeta}(u)}\, G + (-1)^{\eta_{\delta F/\delta\bar{\zeta}}} \frac{\delta F}{\delta\bar{\zeta}(u)}\frac{\delta G}{\delta\zeta(u)} + (-1)^{\eta_F} \frac{\delta F}{\delta\zeta(u)}\frac{\delta G}{\delta\bar{\zeta}(u)}$$

$$+ F\frac{\delta^2 G}{\delta\zeta(u)\delta\bar{\zeta}(u)}$$

Observe that $\eta_{\delta F/\delta\bar{\zeta}} = \eta_F - 1$ since a functional derivative lowers the number of Grassmann variables by one unit. Therefore we can write

$$\mathcal{F} = \frac{\delta^2 F}{\delta\zeta(u)\delta\bar{\zeta}(u)}\, G + F\frac{\delta^2 G}{\delta\zeta(u)\delta\bar{\zeta}(u)} + (-1)^{\eta_F} \left[-\frac{\delta F}{\delta\bar{\zeta}(u)}\frac{\delta G}{\delta\zeta(u)} + \frac{\delta F}{\delta\zeta(u)}\frac{\delta G}{\delta\bar{\zeta}(u)} \right]$$

---

[16]The calculation of the required variational derivatives is analyzed in detail below.

If we look at $\mathcal{D}_1$ and $\mathcal{D}_2$, we see that we can make the identifications

$$F(\bar{\zeta}, \zeta, J) = \left[\frac{1}{i} S_F(0) + \int d^4x \int d^4y\, \bar{\zeta}(x) S_F(x-z)\, S_F(z-y)\zeta(y) \right]$$

$$G(\bar{\zeta}, \zeta, J) = \exp\left\{ -i \int\int d^4x d^4y\, \bar{\zeta}(x) S_F(x-y)\zeta(y) \right\}$$

To start with, we notice that both F and G contain even numbers of Grass-mann variables, and therefore $(-1)^{n_F} = 1$. We thus need to evaluate the following functional derivatives of F

$$\frac{\delta F}{\delta \bar{\zeta}(u)} = \int d^4y\, S_F(u-z)\, S_F(z-y)\zeta(y)$$

$$\frac{\delta F}{\delta \zeta(u)} = -\int d^4x\, \bar{\zeta}(x) S_F(x-z)\, S_F(z-u)$$

$$\frac{\delta^2 F}{\delta \zeta(u)\delta \bar{\zeta}(u)} = S_F(u-z)\, S_F(z-u)$$

and the following functional derivatives of G

$$\frac{\delta G}{\delta \bar{\zeta}(u)} = -i \int d^4y\, S_F(u-y)\zeta(y)\, G$$

$$\frac{\delta G}{\delta \zeta(u)} = i \int d^4x\, \bar{\zeta}(x)\, S_F(x-u)\, G$$

$$\frac{\delta^2 G}{\delta \zeta(u)\delta \bar{\zeta}(u)} = \left[-iS_F(0) + \int d^4x \int d^4y\, \bar{\zeta}(x) S_F(x-u) S_F(u-y)\zeta(y) \right] G$$

Therefore

$$\mathcal{F} = S_F(u-z)\, S_F(z-u)G$$
$$+ \left[\frac{1}{i} S_F(0) + \int d^4x \int d^4y\, \bar{\zeta}(x) S_F(x-z)\, S_F(z-y)\zeta(y) \right]$$
$$\times \left[\frac{1}{i} S_F(0) + \int d^4x \int d^4y\, \bar{\zeta}(x) S_F(x-u) S_F(u-y)\zeta(y) \right] G$$
$$+ i \int d^4x \int d^4y\, \bar{\zeta}(x)\, S_F(x-u) S_F(u-z)\, S_F(z-y)\zeta(y)\, G$$
$$+ i \int d^4x \int d^4y\, \bar{\zeta}(x) S_F(x-z)\, S_F(z-u)\, S_F(u-y)\zeta(y)\, G$$

To evaluate $\mathcal{D}_2$ we also need to calculate the functional derivative with respect to the scalar source

$$\mathcal{S} \equiv \left[\frac{1}{i}\frac{\delta}{\delta J(u)}\right]\left[\int d^4x\, J(x)\Delta_F(x-z)\right]$$

$$\times \exp\left\{\frac{i}{2}\int\int d^4x d^4y\, J(x)\Delta_F(x-y)J(y)\right\}$$

$$= \left[\frac{1}{i}\Delta_F(u-z)+\int d^4x \int d^4y\, J(x)\Delta_F(x-z)\Delta_F(u-y)J(y)\right]$$

$$\times \exp\left\{\frac{i}{2}\int\int d^4x d^4y\, J(x)\Delta_F(x-y)J(y)\right\}$$

We are finally in position to obtain $\mathcal{D}_2$

$$\mathcal{D}_2(\bar\zeta,\zeta,J) = \int d^4z \int d^4u$$

$$\times \left[\frac{1}{i}\Delta_F(u-z)+\int d^4x\int d^4y\, J(x)\Delta_F(x-z)\Delta_F(u-y)J(y)\right]$$

$$\times \left\{S_F(u-z)\,S_F(z-u)\right.$$

$$+\left[\frac{1}{i}S_F(0)+\int d^4x\int d^4y\, \bar\zeta(x)S_F(x-z)S_F(z-y)\zeta(y)\right]$$

$$\times \left[\frac{1}{i}S_F(0)+\int d^4x\int d^4y\, \bar\zeta(x)S_F(x-u)S_F(u-y)\zeta(y)\right]$$

$$+i\int d^4x\int d^4y\, \bar\zeta(x)\,S_F(x-u)S_F(u-z)\,S_F(z-y)\zeta(y)$$

$$+i\left.\int d^4x\int d^4y\, \bar\zeta(x)S_F(x-z)\,S_F(z-u)\,S_F(u-y)\zeta(y)\right\}\tilde W_0(\bar\zeta,\zeta,J)$$

The numerator of $\tilde W(\bar\zeta,\zeta,J)$ expanded to order g^2 then reads

$$\tilde W_{\mathrm{num}}(\bar\zeta,\zeta,J) = \tilde W_0(\bar\zeta,\zeta,J)+ig\mathcal{D}_1(\bar\zeta,\zeta,J)-\frac{g^2}{2}\mathcal{D}_2(\bar\zeta,\zeta,J)$$

(b) The denominator of $\tilde W(\bar\zeta,\zeta,J)$ is obtained setting all the sources to zero;[17]

$$\tilde W_{\mathrm{den}} = \tilde W_{\mathrm{num}}(0,0,0)$$

---

[17]As with QED in Eq. (10.66), the generating functional for QCD in Eq. (10.79) should be completed with $/[\cdots]_{\mathrm{sources}=0}$; the arguments in the text are unaffected.

We therefore need to evaluate $\mathcal{D}_1$ and $\mathcal{D}_2$ letting $\bar{\zeta}, \zeta, J \to 0$

$$\mathcal{D}_1(0,0,0) = 0$$

$$\mathcal{D}_2(0,0,0) = \int d^4z \int d^4u \, \frac{1}{i} \Delta_F(u-z) \left[S_F(u-z) \, S_F(z-u) - S_F^2(0) \right]$$

Hence

$$\tilde{W}_{\text{den}} = 1 - \frac{g^2}{2} \int d^4z \int d^4u \, \frac{1}{i} \Delta_F(u-z) \left[S_F(u-z) \, S_F(z-u) - S_F^2(0) \right]$$

Notice that this expression only contains internal points, over which integrations are carried out.

If we evaluate $\tilde{W}(\bar{\zeta}, \zeta, J)$ up to $O(g^2)$ we have

$$\tilde{W}(\bar{\zeta}, \zeta, J) \approx \tilde{W}_0(\bar{\zeta}, \zeta, J) + ig\mathcal{D}_1(\bar{\zeta}, \zeta, J)$$
$$- \frac{g^2}{2} \left\{ \mathcal{D}_2(\bar{\zeta}, \zeta, J) - \mathcal{D}_2(0,0,0)\tilde{W}_0(\bar{\zeta}, \zeta, J) \right\}$$

Clearly under functional derivation the second term in brackets generates disconnected diagrams, which cancel the disconnected contributions of the companion term (see below).

(c) To evaluate the propagator to $O(g^2)$ one needs to evaluate the functional derivatives contained in

$$iS_F'(x-y) = \left[\frac{1}{i} \frac{\delta}{\delta \bar{\zeta}(x)} \tilde{W}(\bar{\zeta}, \zeta, J) \frac{1}{i} \frac{\delta}{\delta \zeta(y)} \right]_{\zeta = \bar{\zeta} = J = 0}$$

(i) To $O(1)$ we have

$$\left[\frac{1}{i} \frac{\delta}{\delta \bar{\zeta}(x)} \tilde{W}_0(\bar{\zeta}, \zeta, J) \frac{1}{i} \frac{\delta}{\delta \zeta(y)} \right]_{\zeta = \bar{\zeta} = J = 0} = iS_F(x-y)$$

(ii) To $O(g)$ we do not have any contribution since $\mathcal{D}_1$ contains a scalar source J, which is unaffected by the functional derivatives with respect to the Grassmann variables and vanishes when $J \to 0$.

(iii) To $O(g^2)$ there are contributions which we need to evaluate: however, we can set $J = 0$ in the expression for $\mathcal{D}_2$, since only functional derivatives with respect to the Grassmann variables are considered. Similarly, we may disregard all the terms in $\mathcal{D}_2$ which would still contain fermionic sources after the functional derivatives have been performed. For this reason, we may work with

$$\mathcal{D}_2(\bar{\zeta},\zeta,J) \approx \int d^4z \int d^4u \left[\frac{1}{i}\Delta_F(u-z)\right]\left[S_F(u-z)S_F(z-u) - S_F^2(0)\right.$$

$$- 2i\, S_F(0) \int d^4x \int d^4y\, \bar{\zeta}(x)S_F(x-z)S_F(z-y)\zeta(y)$$

$$+ 2i \int d^4x \int d^4y\, \bar{\zeta}(x)S_F(x-u)S_F(u-z)S_F(z-y)\zeta(y)\Bigg]$$

$$\times \exp\left\{-i\int d^4x \int d^4y\, \bar{\zeta}(x)S_F(x-y)\zeta(y)\right\}$$

Therefore, for our purposes, we may consider the generating functional without external scalar sources[18]

$$\tilde{W}(\bar{\zeta},\zeta,0) \approx \tilde{W}_0(\bar{\zeta},\zeta,0) - g^2 \int d^4z \int d^4u\, \Delta_F(u-z)$$

$$\times \left[-S_F(0) \int d^4x \int d^4y\, \bar{\zeta}(x)S_F(x-z)S_F(z-y)\,\zeta(y)\right.$$

$$+ \int d^4x \int d^4y\, \bar{\zeta}(x)S_F(x-u)S_F(u-z)S_F(z-y)\,\zeta(y)\Bigg]$$

$$\times \tilde{W}_0(\bar{\zeta},\zeta,0)$$

where the disconnected contributions originating from the expansion of the numerator of $\tilde{W}$ in powers of g are cancelled by the analogous contributions stemming from the expansion of the denominator.

The fermion propagator up to $O(g^2)$ is therefore

$$iS_F'(x-y) \approx iS_F(x-y) + g^2 \int d^4z \int d^4u\, \Delta_F(u-z)$$

$$\times \left[S_F(x-u)S_F(u-z)\,S_F(z-y) - S_F(x-z)\,S_F(z-y)\,S_F(0)\right] + \cdots$$

and contains only connected diagrams, as anticipated.[19]

(d) We may interpret this result in terms of a set of coordinate-space Feynman rules for this theory:

- Draw all the topologically distinct, connected diagrams containing a given number of external and internal points;

---

[18]We have re-labeled integration variables $z \rightleftharpoons u$ in two of the terms, making use of the fact that $\Delta_F(u-z)$ is symmetric under this interchange.

[19]The two Feynman diagrams for the $O(g^2)$ modification of the fermion propagator are those shown in Figs. 7.5(a) and 12.1(a) in Vol. II.

- Associate a factor $iS_F(x-y)$ to a fermionic line running from the point y to the point x;
- Associate a factor $(1/i)\Delta_F(x-y)$ to a bosonic line connecting the points x and y;
- Associate a factor ig at each vertex;
- Associate a factor of (-1) to each fermionic loop;
- Integrate over the internal points.

These are the Feynman rules derived and stated in Sec. 7.4 of Vol. II.[20]

---

[20] Note there is also a Dirac matrix product along the continuous fermion lines.

PART 4
Problems and Appendices

Appendix A

The Two-Body Problem

Problem A.1 (a) Consider the transformation to relative and center-of-mass coordinates in one dimension

$$x = x_1 - x_2 \qquad ; \ X = \frac{m_1 x_1 + m_2 x_2}{m_1 + m_2}$$

Show that the magnitude of the jacobian for this transformation is unity. Hence conclude that

$$dx \, dX = \left| \frac{\partial(x, X)}{\partial(x_1, x_2)} \right| dx_1 dx_2 = dx_1 dx_2$$

(b) Now establish Eq. (A.9).

Solution to Problem A.1

The jacobian for the transformation to relative and center-of-mass coordinates in one dimension is given by the following determinant

$$\frac{\partial(x, X)}{\partial(x_1, x_2)} = \begin{vmatrix} \partial x/\partial x_1 & \partial x/\partial x_2 \\ \partial X/\partial x_1 & \partial X/\partial x_2 \end{vmatrix}$$

Evaluation of the partial derivatives, and then the determinant, gives

$$\frac{\partial(x, X)}{\partial(x_1, x_2)} = \begin{vmatrix} 1 & -1 \\ m_1/(m_1 + m_2) & m_2/(m_1 + m_2) \end{vmatrix} = \frac{m_1 + m_2}{m_1 + m_2} = 1$$

Hence the volume elements are related by

$$dx \, dX = \left| \frac{\partial(x, X)}{\partial(x_1, x_2)} \right| dx_1 dx_2 = dx_1 dx_2$$

(b) In three dimensions the transformation to relative and center-of-mass coordinates is given by

$$\mathbf{r} = \mathbf{r}_1 - \mathbf{r}_2 \qquad ; \mathbf{R} = \frac{m_1\mathbf{r}_1 + m_2\mathbf{r}_2}{m_1 + m_2}$$

The six-dimensional volume element is

$$d^3r_1 d^3r_2 = dx_1 dx_2 \, dy_1 dy_2 \, dz_1 dz_2$$

Each pair of cartesian components satisfies the relation derived in part (a). Hence

$$d^3r_1 d^3r_2 = dx dX \, dy dY \, dz dZ = d^3r \, d^3R$$

This is Eq. (A.9).

Problem A.2 Show that the commutation relations in Eqs. (A.12) follow from those in Eqs. (A.11).

Solution to Problem A.2

The center-of-mass and relative coordinates, as well as the center-of-mass and relative momenta, are given by

$$\mathbf{R} = \frac{m_1\mathbf{r}_1 + m_2\mathbf{r}_2}{m_1 + m_2} \qquad ; \mathbf{r} = \mathbf{r}_1 - \mathbf{r}_2$$

$$\mathbf{P} = \mathbf{p}_1 + \mathbf{p}_2 \qquad ; \mathbf{p} = \frac{m_2\mathbf{p}_1 - m_1\mathbf{p}_2}{m_1 + m_2}$$

The pairs of vectors $(\mathbf{p}_1, \mathbf{r}_1)$ and $(\mathbf{p}_2, \mathbf{r}_2)$ represent canonically conjugate variables that fulfill the commutation relations in Eqs. (A.11)

$$[p_{1i}, r_{1j}] = \frac{\hbar}{i}\delta_{ij} \qquad ; [p_{2i}, r_{2j}] = \frac{\hbar}{i}\delta_{ij}$$

while variables corresponding to different particles commute

$$[p_{1i}, r_{2j}] = [p_{2i}, r_{1j}] = 0$$

We may now derive the commutation relations among the new variables, using these properties

$$[p_i, r_j] = \frac{m_1}{M}[p_{2i}, r_{2j}] + \frac{m_2}{M}[p_{1i}, r_{1j}] = \frac{\hbar}{i}\delta_{ij}$$

$$[P_i, R_j] = \frac{m_1}{M}[p_{1i}, r_{1j}] + \frac{m_2}{M}[p_{2i}, r_{2j}] = \frac{\hbar}{i}\delta_{ij}$$

and

$$[P_i, r_j] = [p_{1i}, r_{1j}] - [p_{2i}, r_{2j}] = 0$$

$$[p_i, R_j] = \frac{m_1 m_2}{M^2} [p_{1i}, r_{1j}] - \frac{m_1 m_2}{M^2} [p_{2i}, r_{2j}] = 0$$

These are Eqs. (A.12).

Problem A.3 (a) Show $r_1 = R + \mu r/m_1$, and $r_2 = R - \mu r/m_2$; (b) Define

$$\psi(r_1, r_2) = \psi(R + \mu r/m_1, R - \mu r/m_2) \equiv \phi(R, r)$$

Show by explicit calculation of the partial derivatives that

$$\left(\frac{\hbar^2}{2m_1} \nabla_1^2 + \frac{\hbar^2}{2m_2} \nabla_2^2 \right) \psi(r_1, r_2) = \left(\frac{\hbar^2}{2M} \nabla_R^2 + \frac{\hbar^2}{2\mu} \nabla_r^2 \right) \phi(R, r)$$

(c) As an example, take the free-particle solutions

$$\psi(r_1, r_2) = \frac{1}{\sqrt{V}} e^{ik_1 \cdot r_1} \frac{1}{\sqrt{V}} e^{ik_2 \cdot r_2}$$

Determine $\phi(R, r)$, and verify the last relation in part (b).

Solution to Problem A.3

(a) The center-of-mass and relative coordinates, as well as the total mass and reduced mass, are given by

$$R = \frac{m_1 r_1 + m_2 r_2}{m_1 + m_2} \qquad ; r = r_1 - r_2$$

$$M = m_1 + m_2 \qquad ; \mu = \frac{m_1 m_2}{m_1 + m_2}$$

The first two relations are easily inverted to give

$$r_1 = R + \frac{\mu r}{m_1} \qquad ; r_2 = R - \frac{\mu r}{m_2}$$

(b) Use the chain rule of differentiation (twice) on $\psi[r_1(R, r), r_2(R, r)]$. Consider the cartesian component in the x-direction. For the center-of-mass coordinate

$$\frac{\partial}{\partial X} = \frac{\partial x_1}{\partial X}\frac{\partial}{\partial x_1} + \frac{\partial x_2}{\partial X}\frac{\partial}{\partial x_2} = \frac{\partial}{\partial x_1} + \frac{\partial}{\partial x_2}$$

$$\frac{\partial^2}{\partial X^2} = \frac{\partial x_1}{\partial X}\frac{\partial}{\partial x_1}\left(\frac{\partial}{\partial x_1} + \frac{\partial}{\partial x_2}\right) + \frac{\partial x_2}{\partial X}\frac{\partial}{\partial x_2}\left(\frac{\partial}{\partial x_1} + \frac{\partial}{\partial x_2}\right)$$

$$= \frac{\partial^2}{\partial x_1^2} + \frac{\partial^2}{\partial x_2^2} + 2\frac{\partial^2}{\partial x_1 \partial x_2}$$

In addition, for the relative coordinate

$$\frac{\partial}{\partial x} = \frac{\partial x_1}{\partial x}\frac{\partial}{\partial x_1} + \frac{\partial x_2}{\partial x}\frac{\partial}{\partial x_2} = \frac{\mu}{m_1}\frac{\partial}{\partial x_1} - \frac{\mu}{m_2}\frac{\partial}{\partial x_2}$$

$$\frac{\partial^2}{\partial x^2} = \frac{\partial x_1}{\partial X}\frac{\partial}{\partial x_1}\left(\frac{\mu}{m_1}\frac{\partial}{\partial x_1} - \frac{\mu}{m_2}\frac{\partial}{\partial x_2}\right) + \frac{\partial x_2}{\partial X}\frac{\partial}{\partial x_2}\left(\frac{\mu}{m_1}\frac{\partial}{\partial x_1} - \frac{\mu}{m_2}\frac{\partial}{\partial x_2}\right)$$

$$= \frac{\mu^2}{m_1^2}\frac{\partial^2}{\partial x_1^2} + \frac{\mu^2}{m_2^2}\frac{\partial^2}{\partial x_1^2} - 2\frac{\mu^2}{m_1 m_2}\frac{\partial^2}{\partial x_1 \partial x_2}$$

Hence

$$\frac{1}{2M}\frac{\partial^2}{\partial X^2} + \frac{1}{2\mu}\frac{\partial^2}{\partial x^2} = \frac{1}{2m_1}\frac{\partial^2}{\partial x_1^2} + \frac{1}{2m_2}\frac{\partial^2}{\partial x_2^2}$$

Now sum this result over the three cartesian components to arrive at

$$\left(\frac{\hbar^2}{2m_1}\nabla_1^2 + \frac{\hbar^2}{2m_2}\nabla_2^2\right)\psi(\mathbf{r}_1, \mathbf{r}_2) = \left(\frac{\hbar^2}{2M}\nabla_R^2 + \frac{\hbar^2}{2\mu}\nabla_r^2\right)\phi(\mathbf{R}, \mathbf{r})$$

(c) Substitute the relations from part (a) in the free-particle solutions

$$\psi(\mathbf{r}_1, \mathbf{r}_2) = \frac{1}{V}e^{i\mathbf{k}_1 \cdot \mathbf{r}_1}e^{i\mathbf{k}_2 \cdot \mathbf{r}_2} = \frac{1}{V}e^{i(\mathbf{k}_1 + \mathbf{k}_2)\cdot \mathbf{R}}e^{i[(m_2/M)\mathbf{k}_1 - (m_1/M)\mathbf{k}_2]\cdot \mathbf{r}}$$

This allows us to identify $\phi(\mathbf{R}, \mathbf{r})$ as

$$\phi(\mathbf{R}, \mathbf{r}) = \frac{1}{V}e^{i\mathbf{K}\cdot\mathbf{R}}e^{i\mathbf{k}\cdot\mathbf{r}}$$

where $\hbar\mathbf{K}$ and $\hbar\mathbf{k}$ are the total and relative momenta

$$\mathbf{K} = \mathbf{k}_1 + \mathbf{k}_2 \qquad ; \ \mathbf{k} = \frac{m_2\mathbf{k}_1 - m_1\mathbf{k}_2}{m_1 + m_2}$$

These satisfy

$$\frac{\mathbf{K}^2}{2M} + \frac{\mathbf{k}^2}{2\mu} = \frac{\mathbf{k}_1^2}{2m_1} + \frac{\mathbf{k}_2^2}{2m_2}$$

The differential relation in part (b) is then clearly satisfied.

Appendix B

Charged Particle in External Electromagnetic Field

Problem B.1 Verify the vector identity Eq. (B.12).

Solution to Problem B.1

Let $\mathbf{A}$ be the vector potential, with $\mathbf{B} = \boldsymbol{\nabla} \times \mathbf{A}$. Start from

$$[\mathbf{v} \times (\boldsymbol{\nabla} \times \mathbf{A})]_i = \sum_l \sum_m \sum_p \sum_q \varepsilon_{ilm} \varepsilon_{mpq} v_l \nabla_p A_q$$

where ε_{ijk} is the completely antisymmetric Levi-Civita tensor in three dimensions. Use

$$\sum_m \varepsilon_{ilm} \varepsilon_{mpq} = \delta_{ip} \delta_{lq} - \delta_{iq} \delta_{lp}$$

Then

$$[\mathbf{v} \times (\boldsymbol{\nabla} \times \mathbf{A})]_i = \sum_l [v_l \nabla_i A_l - v_l \nabla_l A_i]$$

With the restoration of the vector notation

$$[\mathbf{v} \times (\boldsymbol{\nabla} \times \mathbf{A})]_i = [\mathbf{v} \times \mathbf{B}]_i = \mathbf{v} \cdot \frac{\partial \mathbf{A}}{\partial x_i} - \mathbf{v} \cdot \boldsymbol{\nabla} A_i$$

This is the vector identity in Eq. (B.12).

Bibliography

Abers, E. S., and Lee, B. W., (1973). *Phys. Rep.* **9**, 1

Amore, P., (2014). *Unpublished*

Amore, P., and Walecka, J. D., (2013). *Introduction to Modern Physics: Solutions to Problems*, World Scientific Publishing Company, Singapore

Andraši, A., and Taylor, J. C., (2012). *Ann. of Phys.* **327**, 2591

Bender, C. M., and Wu, T. T., (1969). *Phys. Rev.* **184**, 1231

Bethe, H. A., and Placzek, G., (1937). *Phys. Rev.* **51**, 450

Bethe, H. A., (1949). *Phys. Rev.* **76**, 38

Bjorken, J. D., and Drell, S. D., (1964). *Relativistic Quantum Mechanics*, McGraw-Hill, New York, NY

Bjorken, J. D., and Drell, S. D., (1965). *Relativistic Quantum Fields*, McGraw-Hill, New York, NY

Bloch, F., (1928). *Zeit. für Phys.* **52**, 555

Blatt, J. M., and Weisskopf, V. F., (1952). *Theoretical Nuclear Physics*, John Wiley and Sons, New York, NY

Brack, M., and Bhaduri, R. K., (1997). *Semiclassical Physics*, Addison-Wesley, Reading, MA

Brillouin, L., (1926). *Compt. Rend. de l'Acad. des Sci.* **183**, 24

Cohen-Tannoudji, C., Diu, B., and Laloe, F., (2006). *Quantum Mechanics, 2 vols.*, Wiley-Interscience, New York, NY

Christenson, J. H., Cronin, J. W., Fitch, V. L., and Turlay, R., (1964). *Phys. Rev. Lett.* **13**, 138

Czyż, W., and Maximon, L. C., (1969). *Ann. Phys.* **52**, 59

Dirac, P. A. M., (1947). *The Principles of Quantum Mechanics, 3rd ed.*, Oxford University Press, New York, NY

Donnelly, T. W., Dubach, J., and Walecka, J. D., (1974). *Nucl. Phys.* **A232**, 355

Edmonds, A. R., (1974). *Angular Momentum in Quantum Mechanics*, 3rd printing, Princeton University Press, Princeton, NJ

Faddeev, L. D., and Popov, V. N., (1967). *Phys. Lett.* **25B**, 29

Feshbach, H., (1991). *Theoretical Nuclear Physics Vol. II, Nuclear Reactions*, John Wiley and Sons, New York, NY

Fetter, A. L., and Walecka, J. D., (2003). *Quantum Theory of Many-Particle Systems*, Dover Publications, Mineola, NY; originally published by McGraw-Hill, New York, NY (1971)

Fetter, A. L., and Walecka, J. D., (2003a). *Theoretical Mechanics of Particles and Continua*, Dover Publications, Mineola, NY; originally published by McGraw-Hill, New York, NY (1980)

Feynman, R. P., (1963). *Acta Phys. Polon.* **24**, 697

Feynman, R. P., and Hibbs, A. R., (1965). *Quantum Mechanics and Path Integrals*, McGraw-Hill, New York, NY

Fitch, V. L., and Rainwater, J., (1953). *Phys. Rev.* **92**, 789

Gasiorowicz, S., (2003). *Quantum Physics, 3rd ed.*, John Wiley and Sons, New York, NY

Gell-Mann, M., and Ne'eman, Y., (1964). *The Eightfold Way*, W. A. Benjamin, New York, NY

Glauber, R., (1959). *Lectures in Theoretical Physics, Vol. I*, Interscience, New York, NY

Goldberger, M. L., and Watson, K. M., (2004). *Collision Theory*, Dover Publications, Mineola, NY

Gottfried, K., (1966). *Quantum Mechanics, Vol. I*, W. A. Benjamin, New York, NY

Gottfried, K., and Yan, T.-M., (2004). *Quantum Mechanics: Fundamentals, 2nd ed.*, Springer, New York, NY

Griffiths, D. J., (2004). *Introduction to Quantum Mechanics, 2nd ed.*, Benjamin Cummings, San Francisco, CA

Gross, D. J., and Wilczek, F., (1973). *Phys. Rev. Lett.* **30**, 1343; *Phys. Rev.* **D8**, 3633

Hofstadter, R., (1956). *Rev. Mod. Phys.* **28**, 214

Jackson, J. D., (1998). *Classical Electrodynamics, 3rd ed.*, John Wiley and Sons, New York, NY

Jacob, M., and Wick, G. C., (1959). *Ann. of Phys.* **7**, 404

Koutschan, C., and Zeilberger, D., (2010). *http://astrophysics.fic.uni.lodz.pl/100yrs/pdf/04/076.pdf (viewed May 15, 2010)*

Kramers, H. A., (1926). *Zeit. für Phys.* **39**, 828

Landau, L. D., and Lifshitz, E. M., (1981). *Quantum Mechanics, 3rd ed.*, Butterworth-Heinemann, New York, NY

Lee, T. D., and Yang, C. N., (1956). *Phys. Rev.* **104**, 254

Leith, D. W. G. S., (1974). *Proceedings of the Summer Institute on Particle Physics, SLAC Report No. 179*, ed. M. C. Zipf, Vol. 1, p. 101, Stanford Linear Accelerator, Stanford, CA

Lüders, G., (1954). *Math. Fisik. Medd. Kgl. Danske Aked. Ved.* **28**, 5

Lüders, G., (1957). *Ann. of Phys.* **2**, 1

Merzbacher, E., (1998). *Quantum Mechanics, 3rd ed.*, John Wiley and Sons, New York, NY

Messiah, A., (1999). *Quantum Mechanics, 2 vols.*, Dover Publications, Mineola, NY

Morse, P. M., and Feshbach, F., (1953). *Methods of Theoretical Physics, 2 vols.*, McGraw-Hill, New York, NY

Newton, R., (1982). *Scattering Theory of Waves and Particles, 2nd ed.*, Springer-Verlag, New York, NY

Ohanian, H. C., (1995). *Modern Physics, 2nd ed.*, Prentice-Hall, Upper Saddle River, NJ

Particle Data Group, (2012). *http://pdg.lbl.gov/*

Pekeris, C. L., (1959). *Phys. Rev.* **115**, 1216

Politzer, H. D., (1973). *Phys. Rev. Lett.* **30**, 1346; (1974) *Phys. Rep.* **14**, 129

Sakurai, J. J., and Napolitano, J. J., (2010). *Modern Quantum Mechanics, 2nd ed.*, Addison-Wesley, Reading, MA

Schiff, L. I., (1968). *Quantum Mechanics, 3rd ed.*, McGraw-Hill, New York, NY

Schwinger, J. S., (1947). *Phys. Rev.* **72**, 742

Schwinger, J. S., (1951). *Phys. Rev.* **82**, 914

Serot, B. D., and Walecka, J. D., (1986). *Advances in Nuclear Physics* **16**, eds. J. W. Negele and E. Vogt, Plenum Press, New York, NY

Shankar, R., (1994). *Principles of Quantum Mechanics, 2nd ed.*, Springer, New York, NY

Walecka, J. D., (2001). *Electron Scattering for Nuclear and Nucleon Structure*, Cambridge University Press, Cambridge, UK

Walecka, J. D., (2004). *Theoretical Nuclear and Subnuclear Physics, 2nd ed.*, World Scientific Publishing Company, Singapore; originally published by Oxford University Press, New York, NY (1995)

Walecka, J. D., (2007). *Introduction to General Relativity*, World Scientific Publishing Company, Singapore

Walecka, J. D., (2008). *Introduction to Modern Physics: Theoretical Foundations*, World Scientific Publishing Company, Singapore

Walecka, J. D., (2010). *Advanced Modern Physics: Theoretical Foundations*, World Scientific Publishing Company, Singapore

Walecka, J. D., (2013). *Topics in Modern Physics: Theoretical Foundations*, World Scientific Publishing Company, Singapore

Wentzel, G., (1926). *Zeit. für Phys.* **38**, 518

Whittaker, E. T., and Watson, G. N., (1969). *A Course of Modern Analysis, 4th ed.*, Cambridge University Press, London, UK

Wigner, E. P., and Weisskopf, V. F., (1930). *Zeit. für Phys.* **63**, 54

Wikipedia, (2011). http://en.wikipedia.org/wiki/(topic)

Wu, C. S., Ambler, E., Hayward, R. W., Hoppes, D. D., and Hudson, R. P., (1957). *Phys. Rev.* **105**, 1413

Yang, C. N., and Mills, R. L., (1954). *Phys. Rev.* **96**, 191

Index